The Science of Plants

THE SCIENCE OF PLANTS

Edited by:

Indrachand Chavan

Delve
PUBLISHING

www.delvepublishing.com

The Science of Plants

Indrachand Chavan

Delve Publishing
224 Shoreacres Road
Burlington, ON L7L 2H2
Canada
www.delvepublishing.com
Email: orders@arclereducation.com

ISBN: 978-1-77469-513-5 (Hardcover)

ABOUT THE EDITOR

Dr. Indrachand Chavan (1983) is presently serving as a Principal, Maratha Vidya Prasarak Samaj's KDSPCollege of Agriculture, Nashik. He obtained his B. Sc.(Ag.) in 2006 from Dr. BSKKV, Dapoli, Ratnagiri, M. Sc. (Ag.) in 2009 from Dr. Panjabrao Deshmukh Krishi Vidyapeeth, Akola and Ph. D. Agronomy in 2014 from Sr BSKKV, Dapoli. He started his career as Assistant professor in 2008. His field of specialization is crop husbandry. He has published 01book and 09 research paper in National and international journals of repute. He has participated many National, State Seminar and symposiums. He has delivered Radio talk, Lectures in farmer traning programme.

TABLE OF CONTENTS

CRISPR	Clustered Regularly Interspaced Short Palindromic Repeats
DAD	Diode Array Detecto
DHA	Docosahexaenoic Acid
DNA	Deoxyribonucleic Acid
DNS	Dinitrosalicylic Acid
DTT	Dithiothreitol
E. COLI	Escherichia coli
EACC	Carcinoma Cell Line and Ascites Carcinoma Cell
EC	Electrical Conductivity
EDTA	Ethylenediamine Tetraacetic Acid
ELISA	Enzyme-Linked Immunosorbent Test
EPA	Environmental Protection Agency
EPA	Eicosapentaenoic Acid
EU	European Union
FDA	Food and Drug Administration
FIV	Feline immunodeficiency virus
FT-NIR	Fourier Transform Near Infrared
GA	Gibberellic Acid
GABA	Glutamate and Gamma-Aminobutyric Acid
GBAO	Ground-Based Active Optical
GC	Gas Chromatography
GFP	Green Fluorescent Protein
GFP	Green Fluorescent Protein
GIBS	Genebank Information System
GIS	Geographic Information Systems
GM	Genetically modified
GPS	Global Positioning System
GPX	Glutathione Peroxidase
GS	Genomic Selection
GS	Glutamine Synthetase
GSH	Glutathione
GUS	Glucuronidase
GWAS	Genome-Wide Association
HCL	Hydrogen Chloride

HIV	Human Immunodeficiency Virus
HRAP	High-Rate Algae Pond
HSV	Herpes simplex virus
IPGI	International Peanut Genome Initiative
ISAAA	International Service for the Acquisition of Agri-Biotech Applications
ISSR	Inter Simple Sequence Repeats
LAI	Leaf Area Index
LC	Liquid Chromatography
Luc	luciferase
MAAs	Mycosporine-Like Amino Acids
MAS	Marker-Assisted Selection
MNUGLs	Miscellaneous Neglected and Underutilized Grain Legumes
mQTL	Metabolite Quantitative Trait Locus
mRNA	Messenger RNA
MRSA	Multidrug-Resistant Staphylococcus Aureus
MS	Mass Spectrometry
MZ	Management Zones
N	Nitrogen
NAD	Nicotinamide Adenine Dinucleotide
NBPGR	National Bureau of Plant Genetic Resources
NDVI	Normalized Difference Vegetation Index
NEMS	Nano-Electromechanical Systems
NGS	Next-Generation Sequencing
NIAS	National Institute of Agrobiologically Sciences
NMR	Nuclear Magnetic Resonance
NNIN	N Nutrition Index
NR	Nitrate Reductase
NUE	Nitrogen Use Efficiency
NUS	Neglected and Underutilized Species
OM	Organic Matter
P	Phosphorus
PCA	Principal Component Analysis
PCR	Polymerase Chain Reaction
PEG	Polyethylene Glycol

Ph	Potential of Hydrogen
POME	Palm Oil Mill Effluent
PPy	Polypyrrole
PSI	Photosystem I
PUFA	Polyunsaturated fatty acid
PVX	Potato Virus X
QDs	Quantum Dots
RAPD	Random Amplified Polymorphic DNA
RFLP	Restriction Fragment Length Polymorphism
RGR	Relative Growth Rate
RNA	Ribonucleic Acid
ROS	Reactive Oxygen Species
SDGs	Sustainable Development Goals
SFP	Single Feature Polymorphisms
SNP	Single Nucleotide Polymorphism
SOD	Superoxide Dismutase
SOM	Soil Organic Matter
SoyKB	Soybean Knowledge Base
SPS	Sucrose Phosphate Synthase
SS	Sucrose Synthase
SSCP	Single Strand Such As Complementary Polymorphism
SSR	Microsatellites
STS	Sequence-Tagged Site
TAG	Triacylglycerides
TALEN	Transcription Activator-Like Effector Nucleases
TCA	Tricarboxylic Acid Cycle
TEs	Transposable Elements
TFIID	Transcription Factor II D
TMV	Tobacco Mosaic Virus
tSMS	True Single Molecule Sequencing
UAN	Urea and Ammonium Nitrate
USDA	United States Department of Agriculture
UV	Ultraviolet
WGRS	Whole Genome Re-Sequencing

WRC	World Rice Core Collection
YAC	Yeast Artificial Chromosome
ZnO	Zinc oxide

GLOSSARY

A

Abduction - The act of making a person go somewhere with you, especially using threats or violence.

Adsorbents - An adsorbent is an insoluble material coated by liquid on the surface, including capillaries and pores.

Agronomical- Agronomy is the science and technology of producing and using plants by agriculture for food, fuel, fiber, chemicals, recreation, or land conservation. Agronomy has come to include research in plant genetics, plant physiology, meteorology, and soil science.

Alleles - The term allele denotes the variant of a given gene. In genetics it is normal for genes to show deviations or diversity – all alleles together make up the set of genetic information that defines a gene.

Allergenicity - It is the capacity of a food to cause an allergy. However, this will happen only in genetically predisposed (atopic) individuals, whereas non-atopic individuals can safely consume the same food without any immune, IgE-mediated side effects.

Anion - The ion in an electrolyzed solution that migrates to the anode; broadly: a negatively charged ion.

Antinutrient - A substance that interferes with the utilization of one or more nutrients by the body, as oxalate and phytate, which prevent calcium absorption.

Antioxidants - Antioxidants are compounds that inhibit oxidation, a chemical reaction that can produce free radicals and chain reactions that may damage the cells of organisms. Antioxidants such as thiols or ascorbic acid may act to inhibit these reactions.

Arbitrary – Existing or coming about seemingly at random or by chance or as a capricious and unreasonable act of will.

B

Bioavailability - The ability of a drug or other substance to be absorbed and used by the body.

Biofilm - Biofilm is a complex structure of microbiome having different bacterial colonies or single type of cells in a group; adhere to the surface.

Biopharming - The production and use of transgenic plants and animals genetically engineered to produce pharmaceutical substances for use in humans or animals. It often involves the insertion of gene constructs derived from humans.

C

Canopy - The canopy is the aboveground portion of a plant cropping or crop, formed by the collection of individual plant crowns. In forest ecology, canopy also refers to the upper layer or habitat zone, formed by mature tree crowns and including other biological organisms.

Carotene - Any of several orange or red crystalline hydrocarbon pigments (such as C40H56) that occur in the chromoplasts of plants and in the fatty tissues of plant-eating animals and are convertible to vitamin A.

Catalyzed - Any substance that increases the rate of a reaction without itself being consumed. Enzymes are naturally occurring catalysts responsible for many essential biochemical reactions.

Catastrophe – A sudden event that causes very great trouble or destruction.

Contaminants - A contaminant is either a biological, chemical, physical or radiological substance that becomes harmful to humans or living organisms, when accidentally or deliberately introduced to air, water, soil or food. Contaminants are also known as pollution in environmental science when referring to pollution on a large scale in relation to humans or living organisms.

Crustaceans - A crustacean is an animal with a hard shell and several pairs of legs, which usually lives in water.

Cybridization - The process of protoplast fusion resulting in the development of a cybrid is called as Cybridization.

D

Denitrification - Denitrification is a microbially facilitated process where nitrate is reduced and ultimately produces molecular nitrogen through a series of intermediate gaseous nitrogen oxide products.

Dideoxynucleosides - Dideoxynucleotides are chain-elongating inhibitors of DNA polymerase, used in the Sanger method for DNA sequencing. They are also known as 2',3' because both the 2' and 3' positions on the ribose lack hydroxyl groups, and are abbreviated as ddNTPs.

E

Electrophoresis – It is a laboratory technique used to separate DNA, RNA or protein molecules based on their size and electrical charge.

Enormous – Is marked by extraordinarily great size, number, or degree; especially : exceeding usual bounds or accepted notions.

Erosion- Erosion is the action of surface processes that removes soil, rock, or dissolved material from one location on the Earth's crust, and then transports it to another location where it is deposited.

F

Fodder - Fodder, also called provender, is any agricultural foodstuff used specifically to feed domesticated livestock, such as cattle, rabbits, sheep, horses, chickens and pigs. "Fodder" refers particularly to food given to the animals, rather than that which they forage for themselves.

G

Gene- A gene is a basic unit of heredity and a sequence of nucleotides in DNA that encodes the synthesis of a gene product, either RNA or protein. During gene expression, the DNA is first copied into RNA.

H

Hybridization - The process of producing a plant or animal from two different types of plant or animal

I

Immobilization - The process of stopping something or someone from moving.

L

Leaching - Leaching is the process of a solute becoming detached or extracted from its carrier substance by way of a solvent. Leaching is a naturally occurring process that scientists have adapted for a variety of applications with a variety of methods.

M

Metabolite- A metabolite is an intermediate or end product of metabolism. The term metabolite is usually used for small molecules.

Microfluidics - Microfluidics refers to the behavior, precise control, and manipulation of fluids that are geometrically constrained to a small scale at which surface forces dominate volumetric forces.

N

Nano clay - Nano clays are nanoparticles of layered mineral silicates.

Nodulation – It is the process of forming nodules and especially root nodules containing symbiotic bacteria.

Nutraceutical - A nutraceutical or bioceutical is a pharmaceutical alternative that claims physiological benefits.

P

Phosphorylation - Phosphorylation is the attachment of a phosphate group to a molecule or an ion. This process is common in biology. Protein phosphorylation often activates many enzymes.

Phyto – It is a combining form used like a prefix meaning "plant." It is often used in scientific terms, especially in biology.

Phytol -A hydrophobic alcohol, C20H40O, that occurs esterified as a side chain in the chlorophyll molecule

Phytologist - A biologist specializing in the study of plants.

Phyto nanotechnology – It refers to the synthesis of NMs using plant biomolecules.

Plantibodies - A plantibody is an antibody produced by genetically modified plants.

Polymerases - It is an enzyme that synthesizes long chains of polymers or nucleic acids. DNA polymerase and RNA polymerase are used to assemble DNA and RNA molecules, respectively, by copying a DNA template strand using base-pairing interactions or RNA by half ladder replication.

Prions - Prions are misfolded proteins with the ability to transmit their misfolded shape onto normal variants of the same protein. They characterize several fatal and transmissible neurodegenerative diseases in humans and many other animals.

R

Reflectance - The reflectance of the surface of a material is its effectiveness in reflecting radiant energy. It is the fraction of incident electromagnetic power that is reflected at the boundary.

Resilience – It is the capability of a strained body to recover its size and shape after deformation caused especially by compressive stress.

Ripening - Ripening is a process in fruits that causes them to become more palatable. In general, the fruit becomes sweeter, less green, and softer as it ripens. Even though the acidity of fruit increases as it ripens, the higher acidity level does not make the fruit seem tarter.

S

Sediments - Sediment is a naturally occurring material that is broken down by processes of weathering and erosion, and is subsequently transported by the action of wind, water, or ice or by the force of gravity acting on the particles.

Sonication - Sonication is the act of applying sound energy to agitate particles in a sample, for various purposes such as the extraction of multiple compounds from plants, microalgae and seaweeds.

Spatially - In a way that relates to the position, area, and size of things.

Spectroscopy - Spectroscopy is the general field of study that measures and interprets the electromagnetic spectra that result from the interaction between electromagnetic radiation and matter as a function of the wavelength or frequency of the radiation.

Superweeds - A weed which is extremely resistant to herbicides, especially one created by the transfer of genes from genetically modified crops into wild plants.

T

Tactics - It is the science and art of disposing and maneuvering forces in combat.

Thermostability - It is the quality of a substance to resist irreversible change in its chemical or physical structure, often by resisting decomposition or polymerization, at a high relative temperature.

Topography - Topography is the study of the forms and features of land surfaces.

Transborder -Crossing or extending across a border between two countries.

Tryptophan - Tryptophan is an α-amino acid that is used in the biosynthesis of proteins. Tryptophan contains an α-amino group, an α-carboxylic acid group, and a side chain indole, making it a polar molecule with a non-polar aromatic beta carbon substituent.

Tumor - A swelling of a part of the body, generally without inflammation, caused by an abnormal growth of tissue, whether benign or malignant.

Tundra - A cold, treeless, usually lowland area of far northern regions.

U

Ubiquitous – It means existing or being everywhere, especially at the same time.

V

Virology - Virology is the scientific study of viruses – submicroscopic, parasitic organisms of genetic material contained in a protein coat – and virus-like agents.

Volatilization - Volatilization is the process whereby a dissolved sample is vaporized. In atomic spectroscopy, this is usually a two-step process.

PREFACE

This book introduces the readers to Plant Science. This book sheds light on New Age Molecular Techniques, Transgenic plants, Plant Metabolomics, Plant Phosphate Nutrition and Environmental Challenges, Improving Nitrogen and Phosphorous efficiency for optimal yield and growth, Nanotechnology in Agriculture, Microalgal Biotechnology and Neglected and Underutilized Legume Crops.

The first chapter introduces the readers to Molecular Techniques. This chapter also emphasize the concept of Molecular Marker Analysis, DNA Sequencing, Genetic Linkage Mapping, Marker Assisted Selection, Omic Technologies and Bioinformatic Analyses.

The second chapter takes the readers through an overview of Plant Metabolomics. This chapter will provide highlights on the types, needs, role, advantages, disadvantages and future of transgenic plants. The chapter also explains Plasmid Vectors, Direct and Indirect methods of gene transfers and Gene Constructs for the Production of Transgenic Plants.

Then, the third chapter explains Plant Metabolomics. This chapter also sheds light on different platforms to gather Metabolomic Data, Gene Identification, Plant Metabolomics used for crop improvements, approaches to collate Metabolites and Phenotypic and Genotypic Data. This chapter also explains Metabolomics for the improvement of fruits, legumes, GM crops and Cereal crops.

The fourth chapter takes the readers to Plant Phosphate Nutrition and Environmental Challenges. This chapter introduces the Phosphorus Dynamics in plants, its effects, functions of Phosphorous, factors of Phosphorous availability and environmental factors affecting plant growth.

The fifth chapter introduces the readers to Improving Nitrogen and Phosphorous efficiency for optimal growth and yield. This chapter explains soil and plant analysis, use of tissue analysis and management, spatial variation, fertilizer placement and timing, Leaf area Index and Environmental interaction. The chapter also sheds light on the effect of Nitrogen and Phosphorous application, effects of different rates of Nitrogen Fertilization on crop yield and soil properties.

The sixth chapter throws light on the role of Nanotechnology in Agriculture. This chapter contains the fundamentals, the role of Nanotechnology in Agriculture, Biosensors, Nano-biosensors, Nanotubes etc. This chapter also sheds light on the need for Nano sensors, and new opportunities and challenges of Nanotechnology in Food and Agriculture.

The seventh chapter takes the readers through the concept of Microalgal Biotechnology. The readers are then told about microalgal bioactive compounds, extraction of bioactive compounds, biological activity of microalgae, Biotechnological applications of Microalgae, other applications and products of microalgae and use of Microalgae in Biodiesel production.

The last chapter of this book sheds light on Neglected and Underutilized Legume crops. This chapter also mentions some selected MNUGLs, their genetic potential, their role in sustainable agriculture, etc. The chapter also emphasizes on yield stability in grain legumes, and new technologies for legume improvements and future prospects of MNUGLs.

This book has been designed to suit the knowledge and pursuit of the researcher and scholars and to empower them with various aspects of Plant science, so that they are updated with the information. I hope that the readers find the book explanatory and insightful and that this book is referred by scholars across various fields.

CHAPTER 1

NEW AGE MOLECULAR TECHNIQUES IN PLANT SCIENCE

CONTENTS

Progress in plant science and molecular technologies has grown considerably over the past decade. In certain aspects of breeding and molecular marker-assisted selection processes, Omic strategies, usage of bioinformatics tools, and nanotechnological improvements in agricultural sciences, this book focuses on plant biotechnology applications specializing.

New instructions are needed to adapt their systems to achieve their production goals, and most farmers and breeders can no longer simply turn to the older strategies. On using metabolomics and nanotechnology in agriculture, the book covers new information.

All new data and technology are very important in plant science. The topics are practical and user-friendly in this book. To feel confident about the principles presented in a new generation of molecular plant biotechnology applications, they allow practitioners, students, and academicians with specific background knowledge.

1.1. INTRODUCTION

The study of how plants are used to alter environments, produce food, restore damaged landscapes and well-being, improve community environments, improve human health and provide recreational and practical benefits to the public is known as plant science. To maintain the world around us, plant science is essential. The quality of life for people around the world for generations has been improved by plant scientists.

Early humans undoubtedly observed that plants were greatly affected by the environment. By these early ancestors and passed on to future generations, it envisioned that successful serendipitous plantings, seasonal cropping cycles, and beneficial uses of certain plants were noted.

To provide a more conducive environment for plant growth, agriculturalists have long known the importance of the environment on plant growth, and early farmers were some of the first to manipulate or modify the growing environment.

The understanding of the role of the environment in successful food and fiber production became more of a science as growing plants (and animals) became more specialized, and agriculture continued to evolve.

As fields of advanced study at institutions of higher learning, botany, or the study of plants, developed as a science, investigations on the taxonomy, anatomy, morphology, physiology, and ecology of plants were institutionalized.

Plant sciences are known as the areas of plant study that place major emphasis on the use of plants by humans or as agricultural products (or crops). Plant sciences covers crop production specializations such as Plant pathology, entomology, and integrated pest management, agronomy, horticulture, and forestry.

To a range of research and scientific studies that explore the estimated 410,000 species of land plants found on the earth, plant science is the term given.

Plant science encompasses and involves the observation of plant growth, characteristics, and investigation into the relationship between plants and their environment; it comes from 'Botany' the Greek word βοτάνη(botanē) meaning pasture or fodder, in that. But it is also much, much more.

Figure 1.1. "Plant Science will save the planet."

Source: Image by Flickr

Plants are valuable research objects. Their survival potential under extreme conditions makes them more interesting, and their life spans can reach up to 5000 years. Knowledge about plants has maturated, and deeper research areas have been generated.

The agricultural revolution started after the commercialization of the first transgenic plant. In response to global needs, biotechnological improvements have been rapidly integrated into agricultural technologies.

The plant genetics to thousands of genes, crossing varieties and then selecting the new varieties (and genes) that are desired changed by plant breeding, which is an application. Taking into account the preferences for genetic backgrounds, in this intertwined event, the plant breeder crosses to ensure that the desired traits are gathered in sufficient numbers.

Breeding studies are based on Mendelian genetics. For each cultured plant species, there are several breeding objectives. To suddenly alter these objectives are possible. Therefore, new breeding programs should be adapted.

The breeders must be in close relation with agricultural technologies and the market at this point. The continuous period of mating and selection breeding can be described by the breeders; the only difference is the breeding methods preferred.

To develop plants, biotechnology is often combined with plant breeding in practice. To select the desired plant, genetic markers mapped near genes are responsible in this context for important agricultural features. For plant systems, new-age molecular techniques can be easily adapted. Therefore, a new wide door opened to new possibilities for discoveries in plants.

Releasing the data of plant genomes, it is important to determine the relationship between the interconnected network of genes and gene products. For analysis and interpretation of the results cannot be denied the requirements of new approaches.

The science of plant life and a branch of biology is called botany which is also known as plant science(s), plant biology, orphytology. A scientist who specializes in this field is known as a botanist, plant scientist, orphytologist.

The science of plants is Botany. In order to establish strategies for plant conservation, studying plant classification principles and how they relate to the evolutionary process of the plant is the first step. The molecular properties of plant life play a vital role in plant survival and evolution.

To resist the threats and challenges, they help the plant, such as pollution, human population and activities, and climate change.

To overcome threat effects on plants, some treaties and organizations have established strategies. To understand how these processes are relevant to biodiversity, the methods by which plant species are preserved and recorded for the future are vital. To know the past, present, and future of botanical life is also essential for scientists.

In areas we wouldn't automatically consider it to have applications, botany is surprisingly useful. Aspirin developed as a result of the study of rotting tree bark, and penicillin came from mold; many of our early medicines came from plant extracts.

Especially on plants, as we rely on them to provide carbon sinks amongst other things, the need to study botany is as important today as it ever was, with the effects of climate change expected to have a profound effect on the world around us.

1.2. DIVISION OF PLANT SCIENCE

In a wide variety of disciplines, plant science or Botany today is such a large and varied area and has many important uses that it has evolved into several offshoots and niches. The most notable are as follows, but this list is not exhaustive:

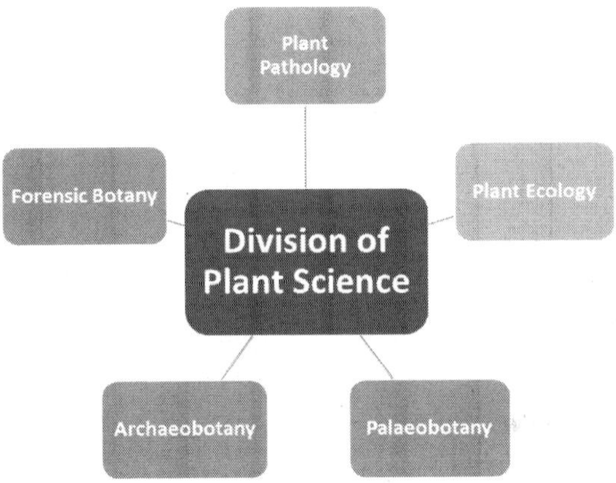

Figure 1.2. Division of Plant Science.

1.2.1. Plant Pathology

In research of plant diseases, it is often tragedy that drives scientific discovery, and the Irish Potato Blight of the 19th century is the one period where most advances - in plant pathology many researchers put their efforts - the reasons should be obvious, people will starve if disease wipes out a large proportion of a staple crop.

Because of the disease that had affected the crops they had come to rely on, the Irish 1840s "Potato Famine" drove the immigration of a large number of Irish to North America largely. The tuber or the roots rot, rendering the whole crop useless, with Potato Blight.

One of the greatest famines anywhere in the western world and has never been seen since caused the mass spread of this disease. As we attempt to adapt crops to the changing climate and keep one step ahead of evolving diseases, today, plant pathology is still a vital component of botany.

How to tackle the many diseases that threaten bananas is an ongoing research problem for plant pathologists. Several types have already gone extinct, and several others remain under threat meant to the limited genetic makeup and the seedless nature of modern cultivars. Understanding disease, as well as the potential problems caused by climate change plant pathology, is vital.

1.2.2. Plant Ecology

Plant ecology is more concerned with how plants interact with their environment; it differs from botany in that - how it reacts to ecological change such as climate change and many more issues of the wider landscape with and in soils, with animal species.

How to breed or otherwise alter our staple plants to survive and adapt to changing environments is vital in a world understanding and adapting to climate change in working out. To grow in what we call "marginal landscapes" as the human population increases and requires more land is also vital to the growing science of GM crops and cross-breeding for hardier crops.

In which plant life exists, plant ecology recognizes eleven major types of environments: coniferous forests, tropical forests, temperate forests, temperate grassland/plain, desert, tropical savannah and, other arid ecosystems, Mediterranean-like regions, tundra, coastal/marine ecology, terrestrial wetlands and, freshwater ecology.

How they interact is as important as each plant's profile in order to understand their evolution because each has its own ecological profile and balanced plant and animal life.

1.2.3. Palaeobotany

The study of extinct plants or fossilized plants recovered from geological strata is known as paleobotany. Just as botanists study modern plants in these taxa, paleobotanists will also study fossilized algae, bacteria, fungi, and lichens. Understanding the changing climate of the past paleobotany has been fundamental.

At any given period of our distant past, it shows us what the ecological makeup of each climate zone was like. How and when the tropical rainforests there developed and what conditions led to their evolution extensive studies in South America have shown precisely.

Such information will tell us about how they might change in the future in light of a changing climate we also expect that. During the Ice Age, paleobotany is also shedding light on nature and the extent of plant species.

1.2.4. Archaeobotany

A botanist will have the tools to enter into the field of studying how people in the past used plants, though most researchers in this area come from an archaeological or anthropological background.

In terms of looking at the spread of irrigation, crop farming, wetland drainage, and other forms of ecological engineering, it can also teach us about how people of the pastexperiencedplants; this discipline can be functional. Some plants have medicinal properties, and others have had - and continue to have – spiritual significance, as we have already discussed.

Sometimes understanding the plant can also help us to understand the beliefs of the past. Archaeobotanists will not just look at the plant remains in the soil but residues in broken pottery and on stone tools, for example. It is vital to separate changes wrought upon a landscape by nature from changes made as a result of human ecological engineering in understanding how environments change.

Archaeobotany has been vital for tracking the spread of the Neolithic Revolution as landscapes undergo wholesale change purely because of human changes; seeds and other plant remains can be great indicators for change, and by extension, in some cases, the spread of the market economy.

1.2.5. Forensic Botany

Forensics has been the area with the most innovation and development; criminal investigations use whatever evidence they can find in the last few decades. What can plants tell us about a crime scene or a body? When we understand the makeup and profile of certain plants, plant evidence can tell us quite a lot, as with pollen and other minute plant remains.

Where a body has been or is presently buried can be said to us by a forensic botanist who will not typically study human remains but will look at plant evidence recovered from bodies or will use criteria that could tell us. It was the case where botany was first employed in a criminal trial; one of the most important cases ever, and criminal historians will say that, is that of the abduction of Charles Lindbergh's son.

To have designed and built such a complex and unique ladder to aid the abduction, a tree expert was called as a witness for the prosecution, and his evidence indicated that the abductor was most likely an expert carpenter; the strong testimony was permitted into the evidence base.

As evidence in a criminal trial, forensic botany, and the use of plants, have come a long way since then, and evidence has been known to determine whether the death was a result of homicide, suicide, or misadventure.

Biology, the study of life, historically was divided into botany, zoology, the scientific study of animals, and the scientific study of plants, which in the past included fungi, bacteria, algae, and lichens. Botany today is more often called plant science or sometimes plant biology.

One group of plants, botanists may specialize in a subject matter field like **ecology** or **genetics** or else study all aspects of (if they study mosses are then called bryologists, for example, or pteridologists if they study ferns). The plant scientists in applied fields like **forestry** or **horticulture** rarely call themselves botanists, but they often study the same processes and structures as botanists do.

In microbial ecology, microbiologistsinvestigate microorganisms such as bacteria and specialize in such aspects.In the structure and function of biological macromolecules, molecular biologists are interested and studying such processes as the biochemical aspects of genetics, while**plant physiologists** analyze metabolic processes such as photosynthesis.

1.3. MOLECULAR TECHNIQUES

In the past centuries and its adverse impact on environmental and socio-economic values, the massive loss of valuable plant species has triggered the conservation of plant resources. For the successful conservation of plant resources and to ensure their sustainable use, appropriate identification and characterization of plant materials are essential.

The past few years have provided easy molecular tools and developed less laborious means for assigning known and unknown plant taxa. These techniques answer many taxonomic questions and new evolutionary, with only phenotypic methods which were not previously possible.

Molecular techniques such as microsatellites and single nucleotide polymorphisms (SNP), DNA barcoding, random amplified polymorphic

DNA (RAPD), and amplified fragment length polymorphism (AFLP) have recently been used for plant diversity studies. Each technique has its own advantages and limitations. In their resolving power to detect genetic differences, these techniques differ in the type of data they generate and their applicability to particular taxonomic levels.

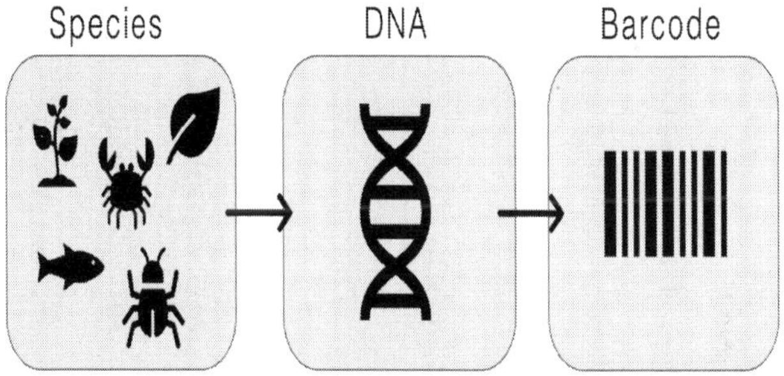

Figure 1.3. Schematic representation of DNA barcoding.

Source: Image by Wikimedia Commons

For DNA fingerprinting and molecular diversity analysis of plant species, this section presents a basic description of different molecular techniques that can be utilized.

Accurate identification of their accession is required by the conservation and sustainable use of plant genetic resources. The practice of species identification techniques changed with the emergence of DNA-based markers has changed.

In the conservation of plant genetic resources with a range of new techniques for easy and reliable identification of plant species, the dramatic advances in molecular genetics over the last few years have provided workers involved.

To study the extent and distribution of variation in species genepools and to answer typical evolutionary and taxonomic questions, many of these techniques have been successfully used.

Properties desirable for ideal DNA markers include highly polymorphic nature, co-dominant inheritance (determination of homozygous and heterozygous states of diploid organisms), selective neutral behavior, a frequent occurrence in the genome, (the DNA sequences of any organism are neutral to environmental conditions or management practices), easy exchange of data between laboratories, easy access (availability), easy and fast assay, high reproducibility.

To classify individuals into nominal genotypic categories and are mostly suitable for intra-species genotypic variation study, at intra-genus and above level, sequencing-based molecular techniques provide better resolution, while frequency data from markers such as RAPD, microsatellites, AFLP and provide the means.

As a basis for making decisions about the conservation of plant resources, this distinction is important to grasp for population studies, particularly when the diversity data are used. For instance, a recent study on Napier grass (*Pennisetum purpureum*) has shown that AFLP is incompatible with RAPD and morphological data;

As a means to resolve the lingering problems regarding the identity of accessions, re-registration of all accessions of Napier grass based on DNA barcoding is suggested. In the conservation of plant resources, the main objective of this section is to provide a basic understanding of the recently developed molecular tools and their potential application.

1.4. MOLECULAR MARKER ANALYSIS

In relation to a gene or a property, a molecular marker can be expressed as a DNA sequence or gene expression product that represents differences in genomic level.

Molecular markers are non-tissue-specific DNA regions that are capable of identifying multiple regions in the genome, reliable, repeatable, standardizable, capable of identifying more than one region in the genome, dominant and co-dominant, and independent of environmental conditions.

As dominant and co-dominant markers, molecular markers are classified. Since dominant markers are not suitable for the identification of heterozygous individuals when related to dominance between alleles being dominated by dominant markers, heterozygous individuals cannot be distinguished from homozygous dominant individuals.

Thus, for any marker at any point, three different individuals (AA, Aa and aa) can be distinguished. In genetic studies conducted by the discovery of the polymerase chain reaction (PCR), the use of molecular marker systems based on this meta-analysis has become more prevalent.

The facilities of the laboratories where the applications will occur, the rapid development of technology and the accompanying needs, the biological properties of the species, and the abundance of markers in the genome have contributed to the development of DNA markers.

Amplified fragment length polymorphism (AFLP), restriction fragment length polymorphism (RFLP), randomly amplified polymorphic DNA markers (RAPD), microsatellites (SSR), sequence labeled sequences (STS), cleaved polymorphic sequence (CAPS), single-strand such as complementary polymorphism (SSCP), amplicon length polymorphism (ALP), expressed sequence tags (EST), interspecific sequence repeat polymorphism (ISSR), and single nucleotide polymorphism (SNP).

As compared to each other, molecular marker techniques have some advantages and disadvantages, but to the purpose of the study and the technical possibilities of the laboratory, their reasons for preference vary according to which the study will be conducted. Because of the high level of polymorphism nowadays, SSR and SNP markers are frequently preferred.

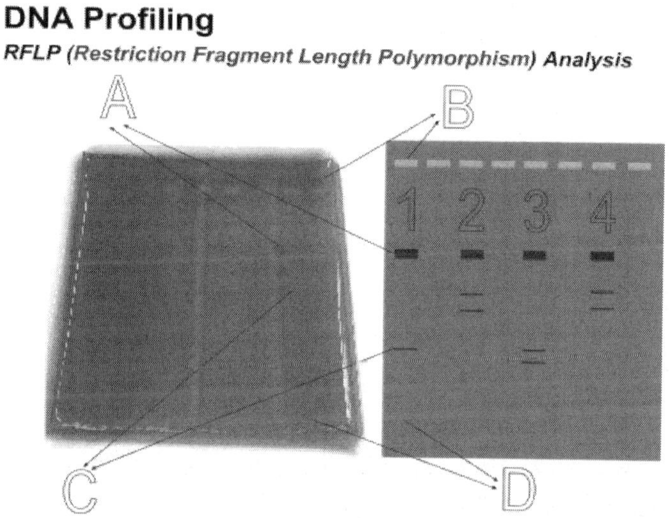

Figure 1.4. DNA Profiling - RFLP Analysis.

Source: Image by Wikimedia Commons

1.5. DNA SEQUENCING

The determination of the order of the nucleotide bases-A (adenine), G (guanine), C (cytosine), and T (thymine) present in a target molecule of DNA is known as DNA sequencing. To obtain DNA sequences within a few days, early work that was developed for the identification and characterization of clinically important bacterial strains has made it possible.

The conventional and next-generation sequencing techniques we describe separately under the following subheadings.

1.5.1. Conventional Sequencing Technique

In automated sequencing analysis, currently, the dye-terminator sequencing technique is the standard method. For the vast majority of sequencing work, the dye-terminator sequencing method, along with automated high-throughput DNA sequence analyzers, is now being used.

Related to dye terminator sequencing and phylogenetic analysis is the basic technique. Labeling of the chain terminator ddNTPs utilized by dye-terminator sequencing utilizes rather than four reactions as in the previously used labeled-primer method, which allows sequencing in a single reaction.

The four dideoxynucleosides chain terminators are labeled with fluorescent dyes in dye-terminator sequencing, each with a different wavelength of fluorescence emission.

The main advantages of this technique are its high accuracy (>98%), robustness, and automation. On the other hand, due to differences in the incorporation of the dye-labeled chain terminators into the DNA fragment, the limitations of this technique include dye effects.

In the electronic DNA sequence trace chromatogram after capillary electrophoresis, such incorporation of dye can result in unequal peak heights and shapes. However, another drawback is its inability to handle long sequences; it can reliably sequence up to approximately 900 nucleotide long DNA fragments in a single reaction.

The advent of new generation sequencers with solid-state chemistry has significantly overcome these problems.

To identify an unknown plant in terms of a known classification is the aim; the current interest is in the DNA barcoding of plants. Using a short DNA sequence from a standard and agreed-upon position in the genome, DNA barcoding is a technique for characterizing species of organisms.

DNA barcode sequences can be obtained reasonably quickly and cheaply, and they are very short relative to the entire genome. Using Basic Local Alignment Search Tool (BLAST) with individual barcodes was obtained with match (99%), followed by trnH-psbA (95%) and then rbcL (75%) the success of species-level assignment of plants.

Use of three-locus DNA barcode resulted in >98% correct identifications of 296 species of woody trees, shrubs, and palms. Recently, as appropriate for the barcoding of plants, a group of plant DNA barcode researchers proposed two chloroplast genes, rbcL and matK, taken together.

Molecular phylogenies in plants are traditionally based on chloroplast DNA (cpDNA) sequence variation. Through the sequencing of coding regions such as rbcL, this approach has proved to be very powerful at the family level.

However, at the genus or species level low evolutionary rate of these sequences limits the power of cpDNA for the assignment. The relationships among closely related taxa have been inferred using non-coding sequences as a consequence.

However, due to the gene flow of cpDNA among closely related taxa, the potential problems triggered the development of new approaches based on nuclear DNA, as well as the lack of phylogenetic resolution. To the sequencing of the ITS (internal transcribed spacer) of 18S–25S nuclear ribosomal DNA corresponded the most common alternative corresponds.

When both fail to resolve phylogenies by cpDNA and ITS sequencing, to solve such difficulties, the AFLP approach has the potential, particularly among closely related species or at the intra-specific level. Therefore, to provide better resolution integration of recently developed barcoding with the following techniques such as RAPD, AFLP, microsatellite and SNP seem.

1.5.2. Next-Generation Sequencing Techniques

On its promise of sequencing DNA at unprecedented speed, a new generation of non-Sanger-based sequencing technologies has been evolving, thereby also having enabled impressive scientific achievements and novel biological applications.

Rather than just short sequences of a single gene, these techniques have made it possible to conduct robust population-genetic studies based on complete genomes. How genotypic variation translates into phenotypic characteristics and rapid progress in genome sequences of various plant species through next-generation sequencing will further extend our understanding.

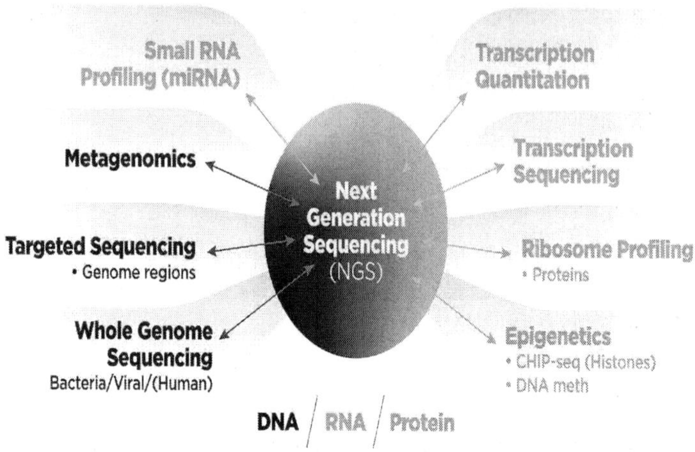

Figure 1.5. Applications of Next-Generation Sequencing Technologies.

Source: Image by NARA & DVIDS Public Domain Archive

To better understand the process of plant evolution, a comparative genomic approach is extraordinarily useful for identifying functional loci related to physiological variation, morphological and geographical, and thus next-generation sequencing technology will be useful.

Next-generation platforms do not rely on Sanger techniques. With the landmark publication of the sequencing-by-synthesis technology developed by 454 Life Sciences based on pyrosequencing, the first of this kind of 2nd generation of sequencing technique appeared in 2005.

By the role of PCR in library preparation, commercial 2nd generation sequencing methods can be distinguished. There are four main platforms; all being amplification-based:

(i) Roche 454 GS FLX
(ii) Illumina Genome Analyzer IIx
(iii) ABI SOLiD 3 Plus System
(iv) Polonator G.007.

The single-molecule sequencing method (also known as 3rd generation or next-next generation) is independent of PC. By Helicos Genetic Analysis System using the technology developed by Braslavskyet al. this mode of sequencing protocol was recently developed.

By Life Technologies and Pacific Biosciences, SMRT technology may appear within one to two years other 3rd generation sequencing systems are being developed. A label-free, electrical, single-molecule, genuinely revolutionary DNA sequencing method Oxford Nanopore Technology has also been developed.

By instead detecting a direct electrical signal, this technique is aimed at obviating the need for amplification or labeling. However, in the developing stage, this technique still exists. The recently developed Helicos 3rd generation high-throughput and low-cost direct single-molecule RNA sequencing method - opened the door for a comprehensive and bias-free understanding of transcriptomes.

Helicos True Single Molecule Sequencing (tSMS) technology significantly increased the speed of sequencing by directly sequencing single molecules of DNA or RNA while also decreasing the cost.

Briefly, on an application, the procedure works by first capturing billions of single molecules of sample DNA -specific proprietary surface within two flow cells as templates for the sequencing-by-synthesis these captured strands serve.

Polymerase and one fluorescently labeled nucleotide (C/G/A/T) are added. On all the templates, the polymerase catalyzes the sequence-specific incorporation of fluorescent nucleotides into nascent complementary strands.

Which removes all free nucleotides after a wash step, the incorporated nucleotides are imaged, and their positions are recorded. In a highly efficient cleavage process, the fluorescent group is removed, leaving behind the incorporated nucleotide. The process continues through each of the other three bases modified from).

Without compromising accuracy or representational bias, scientists at Helicos were able to multiply the system's sample throughput five-fold (from 50 samples to 250 samples per run) using the Helicos DNA Barcoding protocol.

DNA sequencing data from next-generation platforms typically present higher coverage, shorter read lengths, and different error profiles compared with Sanger sequencing data. Especially to cope with the next-generation sequencing data, several software packages have been created.

1.5.3. Strengths of Next-Generation Sequencing

The major strength of next-generation sequencing is that the method can detect abnormalities across the entire genome (whole-genome sequencing only), including substitutions, chromosome inversions/translocations, deletions, duplications, insertions, and copy number changes (gene and exon).

Using less DNA than required for traditional DNA sequencing approaches, a major strength of next-generation sequencing is that it can detect all of those abnormalities. Next-generation sequencing also has a faster turnaround time and is less costly.

1.5.4. Random Amplified Polymorphic DNA (RAPD)

RAPD is a type of PCR reaction in which the segments of DNA that are amplified are random and the primers used are short (8-12 nucleotides); it stands for random amplification of polymorphic DNA.

As the primers will bind somewhere in the sequence, no Knowledge of the DNA sequence for the targeted gene is required, but it is not certain exactly. This technique is based on PCR (Williams et al., 1990; Welsh and McClelland, 1990)

For detecting the polymorphism of genome DNA, it is a convenient, efficient, and sensitive genetic marker. 10 base (GC-rich) oligonucleotide primers of the arbitrary sequence were used in the PCR amplification, and it can get amplification results of several loci.

Since the 1990's to assess intra-specific genetic variation at the nuclear level (Welsh and McClellan, 1990) among them, RAPD markers generated by Polymerase Chain Reaction (PCR) have been widely used. RAPD is a PCR-based technique for identifying genetic variation. Resulting in the amplification of many discrete DNA, RPAD involves the use of a single arbitrary primer in a PCR reaction.

At a very large number of loci, RAPD technology provides a quick and efficient screen for DNA sequence-based polymorphism. The major advantage of RAPD is that it does not require prior DNA sequence information. Moreover, this technique have great diagnostic power and a vast range of potential primers can be used.

By careful selection of primers, reproducible RAPD bands can be found to ensure that only reproducible bands are scored optimization of PCR condition for target species and replication.

For various purposes, which include identification and classification of accessions (Fukuoka et al., 1992), RAPD analysis has been extensively used, genetic diversity analysis (Yu and Paul, 1992; Mackill, 1995; Cao and Oard, 1997), and identification of varieties (Qian et al., 1996).

With single primers of arbitrary nucleotide sequence, RAPD is based on the amplification of genomic DNA. In the absence of specific nucleotide sequence information and the polymorphisms' function, these primers detect polymorphisms as genetic markers and can be used to construct genetic maps.

It is not possible to distinguish whether the amplified DNA segment is heterozygous (two different copies) or homozygous (two identical copies) at a particular locus since most of the RAPD markers are dominant. Co-dominant RAPD markers, in rare cases, are observed as different-sized DNA segments amplified from the same locus may be detected.

The basic technique of RAPD involves

(i) Visualization of RAPD-PCR fragments after ethidium bromide staining under UV light,
(ii) Separation of fragments by gel electrophoresis,
(iii) Addition of a single arbitrary primer,
(iv) Polymerase chain reaction (PCR),
(v) Extraction of highly pure DNA and
(vi) Determination of fragment size compared with a known molecular marker with the help of gel analysis software.

In order to achieve reproducible profiles, it is important to note that the RAPD technique requires maintaining strictly consistent reaction conditions. To reproduce between (and even within) laboratories in practice, band profiles can be difficult if personnel, equipment, or conditions are changed.

The enormous attraction of this technique is that there is no requirement for DNA probes or sequence information for primer design, despite these limitations. No blotting or hybridizing steps are involved in the procedure.

The technique is efficient, quick, and simple and requires only the purchase of relevant chemicals, a thermocycling machine, and agarose gel apparatus, which are available as commercial kits (e.g., Ready-To-Go RAPD analysis beads; GE Healthcare, Buckinghamshire, UK). Another advantage is the requirement for only small amounts of DNA (10–100 ng per reaction)

For detecting genomic variations within and between varieties of sweet potato, the RAPD markers have been used. A total of 160 primers were tested, and eight showed consistent amplified band patterns among the plants with variations within and between varieties of sweet potato.

By RAPD markers and morpho-agronomic characters for a total of 42 accessions of Barberton daisy genetic diversity was evaluated (*Gerbera jamesonii*) employing a set of 12 primer pairs? By using RAPD with the help of 20 random primers, germplasm accession of 80 *Plantago* spp. was studied.

1.5.5. Strength of Random Amplified Polymorphic DNA

RAPDs are quick and easy to assay; that is the main advantage. Only low quantities of template DNA are required because PCR is involved. No sequence data for primer construction are needed since random primers are commercially available. Moreover, throughout the genome, RAPDs have a very high genomic abundance and are randomly distributed.

1.5.6. Amplified Fragment Length Polymorphism (AFLP)

To generate DNA fragments before performing the DNA copying procedure of PCR AFLP, oramplified fragment length polymorphism refers to a method where total genomic DNA extracted from plant tissue is digested with restriction enzymes.

On any single source plant species, it is a generic DNA genotyping technique and can thus be used. To fragment DNA into different lengths, genomic DNA is digested with enzymes, and adaptor sequences (short synthetic segments of DNA) are added to the ends of the fragments using another enzyme called ligase.

To amplify a subset of fragments, each fragment now has a region on each end that can be recognized and bind PCR primers. During the PCR process, by capillary electrophoresis, these fragments are labeled with a fluorescent dye for detection.

By size, labeled DNA fragments are separated by capillary electrophoresis molecular charge, and the fluorescent bar code is captured by a charged coupled device (CCD) camera and analyzed using fragment sizing and genotyping software.

It can be applied to any plant species of interest without a priori knowledge of the genome organization of the species that is the benefit of this technology. The analysis of other forensic DNA techniques also requires a small amount of sample for processing and is similar.

Due to the complexity of the fragment patterns, the limitations to this technology include difficulty in interpreting mixtures, the need for reference databasing for comparisons of the evidentiary plant to representative populations, and the care required to avoid contamination of single-source samples.

For inbred plant species where a majority of the varieties may share common DNA, the complexity of AFLP is highly useful. Due to high levels of DNA fragments and, accordingly, complex band patterns, the more complex the data set, the more likely AFLP will be able to detect a small difference in the DNA.

In restriction enzyme recognition sites, on the basis of the AFLP method, one would expect that different band patterns from related individuals would result from single-base differences. To classify and study many different plant species related to forensic science casework, including *Cannabis* (marijuana), *Olea* (olive), and *Quercus* (oak), AFLP has been successfully used.

The technique involves:

(i) Restriction endonuclease digestion of DNA (enzyme mixture, usually EcoRI + MseI),

(ii) Extraction of highly purified DNA,

(iii) pre-PCR (amplification of the restriction fragments; pre-selective amplification with EcoRI primer + A and MseI primer + C),

(iv) Ligation of adapters (enzyme adapters),

(v) Gel electrophoresis and fragment analysis by automated sequencing machine and

(vi) Selective-PCR with labeled primer pair (Primer + 3 base pairs; forward labeled, reverse unlabeled).

The electrophoretograms can be analyzed using programs like GeneMapper.

Unlike RAPD, the AFLP technique is highly reproducible as it combines restriction digestion and PCR, and AFLP is applicable to all species. However, the automation and recent availability of kits mean that this technology can be brought to a higher level; however, AFLP requires more DNA (300–1000 ng per reaction) and is more technically demanding than RAPD.

To reveal variability when plant species are closely related, nuclear and chloroplast sequences sometimes fail. However, to overcome the hurdles in plant fingerprinting, AFLP distributed throughout the whole genome provides a robust solution.

By Mba and Tohme, a good review on AFLP markers in surveys of plant diversity has been published. Recently, in germplasm collections, several plants such as *Jatropha curcas* and *Rhodiola rosea* have been characterized by AFLP.

To measure the genetic variability within and between natural populations,Teyeret al. have studied the wild populations of*Agave angustifolia* in the desert using AFLP. For phylogenetic analysis and determining the genetic diversity for the conservation of endangered plant species, AFLP markers have been extensively used.

1.5.7. Advantages of Amplified Fragment Length Polymorphism

The advantages of AFLPs lie in their considerable reproducibility, high genomic abundance, the generation of many informative bands per reaction, the fact that no sequence data for primer construction are required, and their wide range of applications.

As clustering in certain genomic regions, such as centromeres, has been reported for some crops, AFLPs may not be totally randomly distributed around the genome. On automatic sequencers, AFLPs can be analyzed, but software problems concerning the scoring of AFLPs are encountered on some systems.

1.5.8. Microsatellites

Polymorphic loci present in DNA that consist of repeating units of one to six base pairs in length are known as microsatellites or simple sequence repeats (SSRs). A $(CA)_n$ repeat, where n is variable among different alleles, is one common example of a microsatellite.

Particularly when the tandem repeats number is 10 or greater, these markers often present high levels of inter-and intra-specific polymorphism. Consisting of two, three, or four nucleotides (di-, tri-, and tetra- nucleotide repeats), the repeated sequence is often simple and can be repeated many times.

Using the unique sequences of flanking regions as primers, microsatellites can be amplified for identification by PCR. Designing PCR primers that are unique to one locus in the genome and that base pair on either side of the repeated portion is the most common way to detect microsatellites.

Therefore, for every individual in the species that produce different sized products, a single pair of PCR primers will work for each of the different lengths of microsatellites. Either by slab gel electrophoresis or capillary gel electrophoresis in an automated sequencer, the PCR products are separated.

Particularly for population analysis, microsatellites have proved to be versatile molecular markers, but they are not without limitations. Primers that flank microsatellite loci are simple and quick to use with the abundance of PCR technology, but the development of functional primers is often a tedious and costly process.

To closely related species, microsatellites developed for particular species can often be applied, but with the increasing genetic distance, the percentage of loci that successfully amplify may decrease. To establish a conservation strategy for endangered plants like*Galiumcatalinense* subspecies *acrispum*, *Calystegia soldanella*, and *Tricyrtis ishiiana* microsatellite technique has recently been used.

1.5.9. Advantages of Microsatellites

The advantages of microsatellites include their high genomic abundance in eukaryotes and their random distribution throughout the genome, and the codominance of alleles. Only low quantities of template DNA are required because the technique is PCR-based.

The reproducibility of microsatellites is high, and analyses do not require high-quality DNA due to the use of long PCR primers. In principle, a single-locus technique, although microsatellite analysis is, multiple microsatellites may be multiplexed during PCR or gel electrophoresis if the size ranges of the alleles of different loci do not overlap.

This significantly decreases the analytical costs. Furthermore, the screening of microsatellite variation can be automated.

1.5.10. Single Nucleotide Polymorphism (SNP)

The development of genomic tools has led to the development of new powerful approaches for mapping complex traits and identifying the genetic basis. Single nucleotide polymorphism (SNP) is presently being used widely for range of genotyping studies.

It is essential to use both traditional breeding techniques and new tools emerging in the field of molecular genetics in parallel with the increase in multidisciplinary studies and the development of technology. The two most used methods in terms of low costs and high performance are microarrays and next-generation sequencing in obtaining data in these tools.

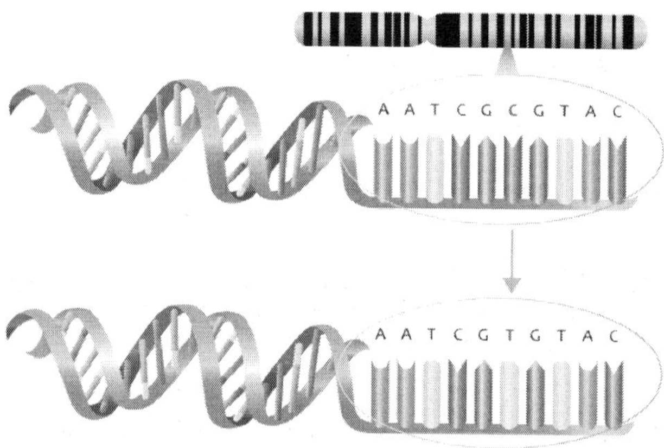

Figure 1.6. Single nucleotide polymorphism substitution mutation diagram.

Source: Image by Wikimedia Commons

In large-scale populations, the selection of SNPs enables the selection of desired lines. To modulate the cultivation program for the determination of the relevant trait, the marker can be used with new-generation technologies than using traditional methods for improvement of the crop more economically.

For fast and cost-effective analysis of germplasm and trait mapping today, conventional plant breeding methods are dependent on SNPs . These differences improve the understanding of genetics that can change the strategy of developing new varieties.

Phenotypic experiments can be attempted faster because the desired trait is under genetic control, and the breeder not only does the early trait selection but also can transmit the desired allele to a large number of populations.

The discovery of interest-specific sequence differences becomes easier as genomes of many species are fully sequenced, including humans, *Arabidopsis*, and rice. These sequences enable the screening of more than

1 million SNPs for each species. As simple genetic markers that can be identified around almost any gene, these polymorphisms can be used.

The usage of SNPs in detecting relationships between allelic forms of a gene and phenotypes creates great potential for characterization of genetic resources, high-resolution genetic map construction, especially common diseases with multifactor genetics, linkage disequilibrium-based association mapping, genetic diversity analysis, genetic diagnostics, cultivar identification, and phylogenetic analysis.

With at least one common (>20% allele frequency) SNP density per kilobase pair, SNPs are frequently found in the genome. It is therefore easy to analyze and mostly biallelic. More importantly, to identify the genes of interest, SNPs allow the combination of candidate gene approach and fusion-based fine mapping.

In terms of illuminating the relationships between varieties, the application of SNPs to genetic diversity is very important. To improve crop plants and protect germplasm allowed plant growers. To identify new alleles in breeding programs, genetic diversity information can then be evaluated.

The results are estimated to extract phylogenetic relationships between species and SNPs that have been applied for several years to assess diversity in specific genes or genomic regions. However, the emergence of new and third-generation technologies can be useful in conserving diversity in domesticated populations and allows the genome to assess SNP-based genetic diversity at scale.

Plant phylogenetic and evolutionary research is traditionally based on sequence genes and hence the knowledge of SNPs. Nuclear and chloroplast gene regions are a rich source of phylogenetic information for evolutionary analysis in plants, which have been preserved for generations.

To explain a wide variety of phylogenetic and evolutionary relationships and inheritance study, the diversity and genotyping of the SNP sequence in these conserved regions are used.

By affecting gene expression and transcriptional and translational promoter activities, SNPs can also be used to discover new genes and their functions. Therefore, in improving agronomic traits, they may be responsible for phenotypic variations between individuals.

Knowing the location of SNP in the genome; it can greatly affect the activity and thermostability level of an enzyme or a similar product if an SNP is present in the coding region. Because some amino acid controls the activity of the expressed regions, sometimes it also depends on the substituted amino acid positions.

To identify various SNPs that can be used for product developments, recent technological advances make it easier. It shows that SNPs in the functional parts of the gene can improve the variety of abiotic and biotic stress tolerance crops by changing expressed regions and controlling the level of biotic and abiotic stress.

1.6. MARKER-ASSISTED SELECTION (MAS)

The process of developing a new, improved variety is accelerated by the Marker-assisted selection (MAS) approach. The MAS is based on the genes that provide the desired character instead of choosing a character (which is a result of many genetic studies). The choice of specific alleles in the marker locus is dependent; this is called the quantitative feature locus.

To Summarize the Theoretical Advantages of MAS:

- Avoid errors that are caused by environmental changes.
- Applicability in a juvenile phase without leaving a character.
- A single plant may be applied here, while the phenotypic selection of some characters requires many seeds or tissues in order to be effective.

Phenotypic selection may be more economical. Although the MAS paternal choice can greatly increase the selection effect of a superior genotype, it does not take the place of sexual recombination and breeding strategies.

Therefore, for improving crop plants, MAS is considered to be an important technique. It is often discussed that phenotypic selection for many characters is faster and cheaper than MAS, and the advantages of MAS may not always be meaningful. Some of the factors that may affect the success of MAS in the negative are as follows:

In terms of technical infrastructure and expertise required for the implementation of MAS, some breeding facilities are inadequate.

- Decreasing the influence of the MAS between the marker and the target QTL.
- The marker must be polymorphic on parents.

MAS is only effective if the selected alleles are more important than the other alleles in the population. The key to the success or failure of every MAS application is this last factor.

MAS is based on the ability to predict the value of alleles. On a number of factors, this prediction depends, but it is essentially an allele, physical environmental conditions that have not yet been tested, and other the behavior of other alleles in existence. For example, the A1 allele at its locus is a positive effect on yield determined by a breeder.

However, in a limited environment and with a limited number of gene sources, this prediction is made. A breeder who crosses a parent with allele A1 and a new parent with the allele A4 and selects the allele A1 as the bound marker will never know that the allele A4 is better than the allele A1, but not in the absence of the allele A1, plants may be susceptible to disease.

For these reasons, from phenotypic selection, MAS should never be applied separately. Rather than applying to phenotype selection, the most successful applications of MAS arise in situations where it is used to improve the traits.

1.7. GENETIC LINKAGE MAPPING

To study diversity and kinship levels, DNA marker technology is used in plants for anlysing genomic information, genomic and physical maps, fingerprint studies, and genomic regions associated with various stress factors.

Identifying similarities or differences among genetic materials is the aim of Fingerprint analysis. The variation in genetic markers represents a variation in genes based on the assumption that fingerprint analysis has been conceived through the use of markers. Markers are widely used to provide information on many genomes at the same time in fingerprint analysis.

The proportion of loci that differs between genetic materials is determined as a result. To select materials to be imported into the plant rehabilitation program, this type of analysis is used, and the breeder has the choice of choosing what he wants from a wider variation with the use of lines with a high variation. For various diagnoses, fingerprint analysis is also used.

Based on genetic markers, fingerprint analysis also has widespread use in forensic studies. Genetic markers are used in genetic variation, evolutionary studies, genomic analysis, the identification of structural changes in chromosomes germplasm analysis, and in the protection of varieties.

Because they allow the whole of the genome to be narrated or done, DNA markers are the most trusted and preferred systems because they are not affected by environment.

Along the chromosome, link maps can determine the position and genetic distance between the markers. By determining how often the marker moves together, a genetic linkage map is formed. On the whole genome, without big gaps, a good genetic map has many markers.

As the rate of use of this information in plant breeding programs increased, the rate of production of genetic maps increased. By DNA markers, both simple and complex inherited genes can be easily identified.

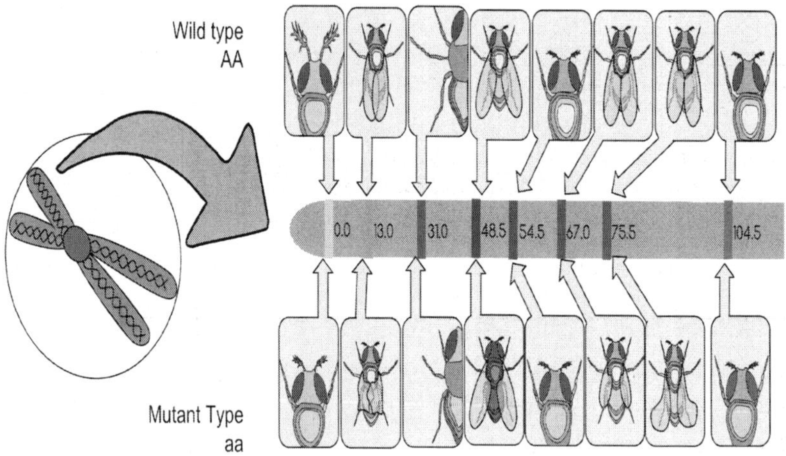

Figure 1.7. Genetic Linkage Mapping.

Source: Image by Wikimedia Commons

To different genotypes in a very short time, thanks to DNA markers provided that genetic maps are first made, many characters (such as resistance to certain diseases) that are simply governed by a single gene have been transferred. The most effective use of molecular markers has been the refinement of quantitative characters possessing complex inheritance and governed by multiple genes.

Many characteristics such as quality, flowering time, plant height, yield and yield components, and endurance against certain diseases and harms are being quantitatively controlled and have considerable prospects for plant breeding trials.

The degree of effect of each locus is different since quantitative characters are governed by multiple loci (QTLs), and it is difficult to determine and transfer in traditional breeding trials because they are highly affected by environmental conditions.

1.8. OMIC TECHNOLOGIES

Between the genes, proteins, and the biochemical pathways, the Omic technologies make the interactions understandable by using several molecular and analytical methods such as bioinformatics and computational analysis methods.

The main focus of Omic technologies is the key traits of interest known as genomics, transcriptomics, and proteomics. The major data sources of omic technologies, such as genomics, transcriptomics, and proteomics forward, have driven improvements in instrumentation and analytical methods.

To use in breeding studies, technological improvements produce genome-scale data. Analysis of the metabolites becomes important in relation to the improvements in analytical methods. Into the responses of the plants against several stress factors, profiling the alterations of the metabolite accumulation provides an insight.

A new omics research field, "metabolomics," was born. Non target metabolome analysis is also useful for evaluating the tissue-specific metabolites and secondary metabolites. In several plant species, it has been reported that significant progress in metabolite quantitative trait locus (mQTL) analysis has been used.

Combining heterogeneous data sets is the main issue in Omic technologies. To describe the different levels of the information of a biological system, high-throughput quantitative Omic data are the best option.

To overcome this problem, computational tools are effective. An integrative comparative analysis of the genomes, analysis of the genome-scale data, phytochemicals, and biosynthetic pathways can be easily and successfully performed.

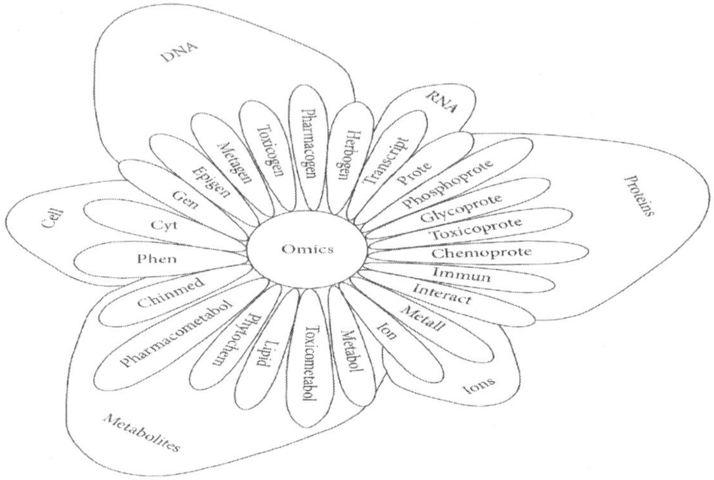

Figure 1.8. Omic approaches and their area of study in systems biology.

Source: Image by Wikimedia Commons

Epigenomics is one of the latest tools to understand gene function regulation in an organism; multi-omics-based systems are demonstrative to understand the pathways or molecules having a role in certain plant functions.

The newest technologies have the opportunity to enable the data to resolve the mechanisms. Epigenomics provides us the ability to define phenotypic variations via DNA-protein interactions, chromatin modifications, and RNA technologies. Also, with next-generation sequencing (NGS) technologies, the usage of chromatin immunoprecipitation (ChIP) techniques can gain epigenomic data from plant species.

1.9. BIOINFORMATIC ANALYSES

By new-age molecular biology techniques, genomic information obtained is required to be stored, organized, and analyzed. Bioinformatics methods have progressed rapidly and exchanged the status of the research. For the processing of largescale data in detail, the use of bioinformatics tools is crucial.

To process and analyze plant genomics data is an important point. The main challenge is NGS technology. The need for tools is obvious in recent years and the increase in the number of sequenced plant genomes.

The heterogeneous nature of the plants and innovative bioinformatics tools have become mandatory. The German Federal ex situ Gene Bank of Agricultural and Horticultural Crops (GCBN), GIBS (Genebank information system), EURISCO, LAILAPS, PlantsDB, IPK blast server, and Plabi PD are recent platforms for plant genomic resources.

In-plant biotechnology studies, the integrative improvements of multiple Omic technologies and computational tools are helpful. To enable the network between different fields of life sciences, interdisciplinary collaborations are important.

For the researchers working on plant biotechnology to provide new insights into agricultural problems, this must be the most important mission. To solve the upcoming problems and define the requirements of plant breeders, otherwise, it will be a big challenge.

1.10. CONCLUSION

In-plant biology, the utility of metabolomic analysis has by now been well demonstrated, yielding valuable insights into mechanisms of biochemical regulation of growth, gene function, nutrient and environmental stress, and development among others.

However, these represent only a tiny fraction of the potential. We should expect even greater information content and improved biological insights as to the technology itself advances. In a manner that yields high quality, consistent, comparable, and timely results, the challenge is to make this technically difficult technology accessible to the entire plant biology community.

In our understanding of metabolic regulation in plant biology, the wide adoption of metabolic profiling in plants will eventually create synergistic payoffs.

REFERENCES

1. Siegel, J., J. Saukko, P. and M. Houck, M., 2013. *Encyclopedia of ForensicSciences | ScienceDirect.* [online] Sciencedirect.com.Available at: <https://www.sciencedirect.com/referencework/9780123821669/encyclopedia-of-forensic-sciences#book-info> [Accessed 2 June 2022].
2. Environmentalscience.org. n.d. *Botany: The Original Natural Science | EnvironmentalScience.org.* [online] Available at: <https://www.environmentalscience.org/botany> [Accessed 2 June 2022].
3. Gonçalves-Vidigal, M. and Rubiano, L., 2011. Development and application of microsatellites in plant breeding. *Crop Breeding and Applied Biotechnology*, [online] 11(spe), pp.66-72. Available at: <https://www.scielo.br/j/cbab/a/dW4DCvLMV47KG8pGjrVjLGs/?lang=en> [Accessed 2 June 2022].
4. HEARNE, C., GHOSH, S. and TODD, J., 1992. Microsatellites for linkage analysis of genetic traits. *Trends in Genetics*, [online] 8(8), pp.288-294. Available at: <https://www.wur.nl/en/show/Microsatellites.htm#:~:text=The%20strengths%20of%20microsatellites%20include,of%20template%20DNA%20are%20required> [Accessed 2 June 2022].
5. Kumari, N. and Kumar Thakur, S., 2013. *Randomly Amplified Polymorphic Dna-a Brief Review.* [ebook] https://thescipub.com. Available at: <https://thescipub.com/pdf/ajavsp.2014.6.13.pdf> [Accessed 2 June 2022].
6. Morgil, H., Can Gercek, Y. and Tulum, I., 2020. Single Nucleotide Polymorphisms (SNPs) in Plant Genetics and Breeding. *The Recent Topics in Genetic Polymorphisms*, [online] Available at: <https://www.intechopen.com/chapters/71577> [Accessed 2 June 2022].
7. R. Decoteau, D., n.d. *Principles of Plant Science.* [ebook] http://www.personal.psu.edu/. Available at: <http://www.personal.psu.edu/drd10/Site/Publications_files/PlantScienceTeachersManual.pdf> [Accessed 2 June 2022].
8. Williams, J., Kubelik, A., Livak, K., Rafalski, J. and Tingey, S., 1990. DNA polymorphisms amplified by arbitrary primers are useful as genetic markers. *Nucleic Acids Research*, [online] 18(22), pp.6531-6535. Available at: <https://www.wur.nl/en/show/Random-Amplified-Polymorphic-DNA-RAPD.htm> [Accessed 2 June 2022].

CHAPTER 2

TRANSGENIC PLANTS: GENE CONSTRUCTS, VECTOR AND TRANSFORMATION METHOD

CONTENTS

Transgenic plants are those that have one or more genes from another species inserted into their genome by genetic engineering procedures. The biolistic method and *Agrobacterium tumefaciens*-driven transformation are two techniques for gene transfer.

Transgenic plants are those whose DNA has been altered via the use of genetic engineering techniques. The goal is to impart a new characteristic to the plant that does not exist in the species usually. The transgene is then added gene sequence, which may be from an unrelated plant or a completely other species.

2.1. INTRODUCTION

The goal of introducing a new gene combination into a plant is to make it as helpful and productive as feasible. This technique has benefits such as increased shelf life, greater yield, higher quality, insect resistance, heat, cold, and drought tolerance; they express foreign proteins having industrial and medicinal utility in this manner.

Plants containing vaccinations or antibodies (Plantibodies) are particularly challenging since plants are devoid of human diseases, lowering screening expenses for viruses and bacterial toxins.

In 1983, the first transgenic organism was produced. Many recombinant proteins have now been expressed in a variety of major agronomic plant species, including tobacco, corn, tomato, potato, banana, alfalfa, and canola. Tobacco plants were commonly employed, but potatoes and bananas are also being studied for human vaccinations.

For several years, humans have genetically modified plant crops through traditional breeding. Until recently, the only means to add phenotypic features of interest, which are dictated by genes, to an individual plant or species was through breeding. Through breeding and selection, the desirable phenotypic qualities are passed down to the descendants.

Nevertheless, such traditional genetic manipulation approaches have several limitations, such as the sexual barrier between species, phylogenetic isolation barriers between and within genetic groupings, a smaller gene pool, and linkage drag. All of these disadvantages, in addition to the considerable period necessary for desirable qualities to be conveyed, are disadvantages.

Lately, molecular biology, cell cultures, and gene transfer techniques have been coupled to provide a potent tool for introducing new traits into a specific plant. As a consequence, scientists may transfer genes from animals, microbes, or various plant species into the genome of a recipient plant in a regulated, fertilization-free way.

As a result, sexual barriers between species are erased, as are barriers between organisms in the evolutionary realms of eubacteria, archaea, and Eukarya.

Plant genetic transformation is the deliberate introduction of particular regions of DNA into the genome of a recipient plant, eliminating introduction by fertilization or hybridization. Other non-sexual gene transfer strategies for plants exist, such as somatic hybridization, which involves fusing two separate protoplasts and combining the genes.

The genomes of the nucleus and the cytoplasm. Cybridization, which happens by cytoplasmic fusion, is another non-sexual transmission technique. However, only genetic transformation allows for the insertion and integration of particular DNA pieces.

Comprehensive research in fundamental molecular and cellular biology is required to find and isolate the gene responsible for the trait of interest in the donor organism in order to implant a desired gene into the host. Only the discovery of plant tissue culture techniques enabled genetic change in plants.

Foreign DNA must first be introduced into a plant cell or plant tissue for the transformation process to be effective; a transgenic plant is then produced from the transformed cell. The totipotency phenomenon, which is the capacity of individual plant cells to develop into completely new plants under favorable conditions and in the presence of growth regulators and nutrients, enables the procedure.

2.2. TYPES OF TRANSGENIC PLANTS

Transgenic plants contain genes from those other species introduced into them. The implanted genes might come from the same kingdom (plant to plant) or from different kingdoms (bacteria to plant). In many circumstances, the inserted DNA must be somewhat changed in order to express appropriately and efficiently in the host organism.

Transgenic plants are utilized to express proteins such as *Bacillus thuringiensis* cry toxins, herbicide resistance genes, and vaccine antigens.

Cisgenic plants are made up of genes occurring in the same or a closely related species, where traditional plant breeding may take place. Some breeders and scientists feel that cisgenic alteration is beneficial for plants that are difficult to crossbreed using traditional methods (such as potatoes). Plants classified as cisgenic should not face the same degree of legal scrutiny as other genetically modified organisms.

Figure 2.1. Cisgenic potatoes plant after exposure to potato blight.

Source: Image by Wikipedia

2.3. STEPS INVOLVED IN THE PRODUCTION OF TRANSGENIC PLANTS:

There are only a few stages required in the development of transgenic plants. These are just as follows:

- Gene Identification, Isolation, and Cloning for Agriculturally Important Traits
- Creating a Gene Construct for Insertion
- Using the Gene to Transform Target Plants Transgenic Plant Tissue/Cell Construct Selection
- Plant Regeneration in Transgenic Plants

These would be the stages that are included in the generation of transgenic plants. Such steps are used to make transgenic plants.

2.4. NEED FOR TRANSGENIC PLANTS

To eliminate hunger and malnutrition and to meet the demands of an economically ascending population in the future century, global food production must be quadrupled by 2030-2050. Furthermore, this growth must be achieved by increasing productivity per unit area while making sustainable use of natural resources such as water, nutrients, and energy.

Plant breeding has been effective in overcoming these issues in several crops. Plant breeders' goal is to create a mixture of novel genes (those responsible for expressing novel features) in one crop plant that will make it as helpful and productive as feasible.

Depending on where and for what purpose the plant is produced, desirable genes may deliver expected qualities such as increased yield or enhanced quality, insect resistance, disease resistance, and tolerance to heat, cold, and drought tolerance, for example.

Even so, trying to combine the best genes in one plant is a time-consuming and hard task, particularly since conventional fertilization has indeed been restricted to artificial crossing plants within the same species, both with closely related species to bring genetic variations together and by the available genetic pool.

It is often impossible to incorporate novel qualities into desirable crops since all advantageous novel traits are never accessible in a specific crop species or even closely related species.

Sexual compatibility between the two parents limits future plant breeding. Because genes for advantageous qualities are not always found in a crop species or its sexually compatible wild relatives, traditional breeding cannot produce plants with the novel, desirable traits taken from other species.

A gene for increasing protein content in soybean, for example, could not be transferred to a completely another crop, such as corn, using conventinal breeding methods.

Transgenic technology has revolutionized agriculture by allowing plant breeders to create more useful and productive crop types by introducing particular traits (genes) from a diverse variety of living sources, not simply those present within the crop species itself. *Bacillus thuringiensis* cry genes (commonly known as Bt genes) were discovered to express proteinaceous toxins inside bacterial cells.

Such poisons can kill certain insect pests such as *Lepidoptera*, *Diptera*, and *Coleoptera* without harming other beneficial insects such as silkworms and butterflies.

Many Bt-transgenic crops (brinjal, cauliflower, cabbage, canola, corn, cotton, eggplant, maize, potato, tobacco, tomato, rice, soybean, and so on) have been created and marketed at the national and international levels using the cry gene. Transgenic development is a very target-oriented strategy with little doubt about the traits transmitted.

2.5. APPLICATIONS OF TRANSGENIC PLANTS

There are two primary areas of use for transgenic plants.

These are some examples:

2.5.1. In Agriculture

The agricultural industry is among the key sectors of agriculture because these approaches include modifying the genetic information of plants. Since 1982, the method has been applied to alter the properties of several plant species spanning from maize and barley to onion and tomato.

Gene transfer is utilized for a variety of purposes, including improving yields (crops and animal feeds, for example) and raising plant/crop tolerance to various biological and environmental conditions.

2.5.2. In Biopharming/Pharmaceutical Industry

Besides the agriculture sector, gene transfer in plants has various uses in the pharmaceutical business. Gene transfer is being used to mass-produce proteins and compounds that plants do not naturally make.

Transgenic plants are thought to become more cost-effective than other cells used in biopharmaceuticals. Furthermore, they have the equipment in place to change molecules into specific structures that can perform the necessary biological activities.

Since about the early 1990s, biotech businesses have advocated employing food and feed crops as small industries for creating medicinal proteins and chemical products which they do not generate naturally.

Such an approach, known as "biopharming," involves inserting foreign genes coding for medically relevant proteins into plant cells, including therapeutic proteins, monoclonal antibodies, and vaccinations. Nevertheless, the FDA has yet to approve a single medication manufactured using this procedure.

Inserting the gene for the desired protein into the DNA of chloroplasts, membrane-bound organelles containing chlorophyll, is one way to biopharming. Chloroplasts contain their own circular set of genes that differs from the main genome in the nucleus of the cell.

Each cell in the leaves of higher plants includes up to 100 chloroplasts, each of which carries up to 100 copies of the genome. By adding a transgene into the chloroplast genome, it is possible to significantly increase the gene and create enormous amounts of the associated protein.

Biopharmaceutical production in transgenic plants may be a more cost-effective alternative to altered bacteria or mammalian cell culture. Plant cells have the biochemical machinery needed to fold complex proteins and undergo post-translational changes (such as glycosylation and the addition of sugar molecules) essential for complete biological activity, which is one benefit of biopharming.

Furthermore, unlike mammalian cells, plants do not contain retroviruses or other infectious agents that lead to diseases in humans (such as prion diseases).

One approach to biopharming is to insert the gene for the desired protein into the DNA of chloroplasts, membrane-bound organelles containing chlorophyll. Chloroplasts have their own circular set of genes that is distinct from the main genome in the cell nucleus.

In the leaves of higher plants, each cell has as many as 100 chloroplasts, each of which contains up to 100 copies of the genome. Thus, by inserting a transgene into the chloroplast genome, one can greatly amplify the gene and produce large amounts of the corresponding protein.

Production of biopharmaceuticals in transgenic plants may offer a cost-effective alternative to using engineered bacteria or mammalian cell culture.

One advantage of biopharming is that plant cells possess the biochemical machinery needed to fold complex proteins and to perform the post-translational modifications (such as glycosylation and the addition of sugar molecules) required for full biological activity. Moreover, unlike mammalian cells, plants do not contain retroviruses and other infectious agents (such as prions) that cause disease in humans.

2.6. ROLE OF TRANSGENIC PLANTS AS BIOREACTORS OR BIOFACTORIES

Plant cells, like so many other cells, have biological machinery that is engaged in a variety of functions. The genetic material stores important information (blueprint) essential for these activities.

Researchers may now make use of these processes to manufacture diverse proteins such as vaccines and antibodies by adding additional information in the form of new genes.

While bacteria and other types of cells have been employed as bioreactors, transgenic plants have gained popularity in recent years due to their lower cost and ability to undergo post-translational modifications necessary in the creation of complex proteins.

2.7. ADVANTAGES OF TRANSGENIC PLANTS

As previously stated, the methods and technology utilized to manufacture transgenic plants are employed in a variety of industries/sectors. This appears to be because of the way that transgenic plants have several benefits over native ones.

Some of the main benefits of transgenic plants include:

Trasgenic plants resistant to various Biotic and Abiotic Stress

Biotic stressors would be those caused by other creatures in nature (bacteria, viruses, fungi, etc.), whereas abiotic stresses are those caused by the ambient circumstances in which a certain plant is cultivated.

Researchers have made various types of plants/crops resistant to certain stresses by inserting particular genes. Farmers are able to avoid the high expenses associated with plant/crop damage caused by biotic and abiotic stresses as a result of this approach.

2.7.1. Increased Yields

Another advantage of transgenic plants is that they produce higher yields than native plants. In general, there are two basic reasons why these plants produce more. The first reason is that researchers may now insert genes that can alter plants, leading to better yields than wild plants/crops.

Furthermore, by inserting genes that allow the plant to endure biotic and abiotic stresses, the plant may survive a variety of environmental circumstances that would normally cause considerable harm to the plant.

2.7.2. High Quality of Yields

Among the most significant advantages of transgenic plants is that they yield higher-quality crops. As previously said, there are several stresses in nature that might impact plant development and yields. Furthermore, several pesticides used to manage insects, pests, and weeds, among other things, have been demonstrated to harm sensitive plants.

Researchers have been able to add new traits to several types of plants using gene transfer, allowing them to tolerate these pressures. As a result, because they are not considerably impacted by elements that affect other plants and their yields, these plants produce high-quality crops.

The primary benefits of a transgenic plant or genetically modified crop include increased production, resistance to diseases and pests, and the ability to flourish under harsh circumstances. Gene technology has a key role in increasing the production of food, fiber, cereals, and vegetable crops by improving output. Controlling losses caused by numerous insects and diseases result in an increase in yield.

Disease and insect resistance were major factors in gene technology. Insects infect agriculture plants, and pesticides are used to control the insects. However, insects have no effect on transgenic plants or genetically engineered crops.

One of the primary benefits of transgenic plants is improved quality. Gene technology has aided in the improvement of all three levels of quality in various crops. It can create more in a less amount of space. Because of its drastically enhanced yields, it can feed a fast-growing population.

It decreases the need for pesticides and insecticides during cultivation, which might be beneficial to the food supply. These are the pros and drawbacks of transgenic plants or genetically engineered crops which can help you comprehend its benefits.

2.8. DISADVANTAGES OF TRANSGENIC PLANTS

2.8.1. Cross-Contamination

Pollen from genetically engineered plants has been polluted. This might lead to superweeds with the same resistant qualities as crops.

2.8.2. Antibiotic Resistance

Because of its built-in disease-control systems, genetically engineered crops contain mutant traits with enhanced antibacterial activities. This can lead to superbugs, which can disrupt the healthcare system.

2.8.3. Carcinogen Exposure Risk

According to studies, crops that are resistant to commercial pesticides significantly increase the risk of cancer development in rats. Genetically engineered crops can also offer major health risks.

The primary drawbacks of transgenic plants are allergic responses, the creation of super-pests, and biodiversity loss.

It raises the expense of cultivation and makes farming more prone to marketization, which results in unjust gains.

Transgenic crops imperil not just farmers but also commerce and the environment. It has been physiologically changed. As a result, biotech foods may endanger human health.

Excessive production of genetically engineered crops will eventually render them useless since the pests that these poisons are designed to repel may develop resistance to them. These are the downsides and disadvantages of transgenic plants or genetically engineered crops. This will help you understand that everything in the world has its own set of advantages and disadvantages.

For many years, the use of transgenic crops was a source of contention. Many concerns have been expressed, and they may be divided into two groups.

Issues have been expressed concerning the potential impact of genetically engineered materials on human health. Transgenic crops, for example, have been linked to allergies in certain people, while it is unclear if transgenic crops are to blame.

Moreover, antibiotic resistance genes inserted into these crops have been linked to antibiotic resistance, resulting in superbugs that cannot be destroyed with antibiotics. The populace is concerned about swallowing DNA from another source, such as a virus or bacteria.

Concerns have been raised concerning whether transgenic crops harm the natural ecosystem. One example is pollen from transgenic corn, which has the potential to destroy Monarch butterfly larvae.

It has been demonstrated that hybrid corn pollen contains a bacterial poison that is distributed over 60 meters by the wind. In this region, maize pollen is deposited on other plants near cornfields, where it might be consumed by non-target creatures such as the monarch butterfly, killing species.

2.9. FUTURE OF TRANSGENIC PLANTS

Though genetically modified crops have the potential to alleviate global food shortages, their practicality for production remains under doubt. The increased cultivation of GM crops to eradicate hunger has unintended consequences for the environment and human health. The topic remains contentious, and the future of genetically engineered crops appears unclear.

The commercial success of transgenic crops from 1994 to 2002 proved that the adoption of transgenic crops for commercial production in farmer's fields would result in major advantages.

The following are significant advantages:

- Improved and more efficient weed control,
- Reduced losses due to insect pests and viruses and reduced need for insecticide,
- Reduced post-harvest losses due to improved shelf life and marketing flexibility (tomato) due to storage pest resistance.
- Increased nutritional quality (oil in canola) and
- More effective hybrid seed production.

The above will not only contribute to a more sustainable food security system but also to a safer environment due to reduced insecticide and pesticide use. This will necessitate a response from the seed industry to address this evolving circumstance by providing farmers with seeds for these improved crops. Developing nations will have to create protocols and policies for the sale and use of these transgenic crops.

Transgenic crops will be utilized in the future not just for better agronomic features but as well as for food production, medicines (including edible vaccinations), and specialty chemicals. Transgenic rubber trees have also been developed for a range of applications. As a result, the future of transgenic crops is bright and hopeful.

The market for these crops is estimated to reach $3 billion in 2000 and $6 billion in 2005. These objectives will be met via continuous efforts in both developed and developing countries. The general population and farmers must adapt to this changing environment.

To reap the advantages of these transgenic crops, which will be produced in vast quantities in the coming decade, the public and commercial sectors will have to play a big role.

2.10. GENE CONSTRUCTS FOR THE PRODUCTION OF TRANSGENIC PLANTS

A promoter region, gene coding region, and terminator/stop region comprise a basic functioning gene construct. Depending on the nature of the research, various gene constructs may also incorporate specific sequences such as an enhancer, silencer, or reporter sequence.

Transgene construction is always the first step in plant transformation. Other than the incorporation of the gene of interest and selectable markers, transgene constructs are essentially identical. A suitable gene construct is essential for the successful production of an optimum transgenic strain.

The difficulty of a gene construct varies depending on the insertion of interest. The construct can be designed to be inserted randomly into the animal's genome, a process known as transgenesis by addition, or it can be designed to be inserted into the genome at a specific targeted site, into the correct position of a determined chromosome, a process known as transgenesis by homologous recombination.

In all circumstances, the construct must be flawless, having components that govern gene products such as a promoter, a transcription initiation site, a polyadenylation site, and a transcription termination site.

Such that, the data being introduced into the receptor's genome does have a start, middle, and end, preventing difficulties with unregulated expression in the host cell. To achieve the goals of the suggested genetic modification research, a wide range of machinery is now available to control the expression of transgenic organisms in a ubiquitous or conditional way.

2.10.1. A Typical Plant Gene

A specific plant gene includes regulatory and structural genes. Regulatory genes are normally found 5′ upstream of a gene and have their own promoter, enhancer, or silencer region. Structural genes, on the other hand, start with a catabolite activator protein (cap) site and continue with a leader sequence, starting codon, exons, introns, terminator, and polyadenylation site (poly-A tail).

These components are in charge of DNA transcription. The transcribed pre-mRNA is subsequently spliced, yielding mature mRNA without introns (non-coding region). This mature mRNA is transferred to the cytoplasm for translation, which is triggered by ribosomal subunit binding to the promoter.

The ribosome subsequently moves downstream to the next codon, producing a peptide chain with the aid of tRNAs, and terminates when it hits the chain terminator (stop codon, TAA/TAG).

2.10.2. Promoters/Enhancers

The promoter domain is usually found 5′ upstream of a gene. Promoters are well-known for their role in regulating gene expression, which may be compared to an on/off switch. Transcription factors identify the promoter sequence during DNA transcription.

These transcription factors bind to the promoter's consensus area and recruit the RNA polymerase. The formation of the RNA polymerase transcription complex signals the start of DNA transcription.

Constitutive promoters, tissue-specific promoters, and inducible promoters are the three primary types of promoters. Constitutive promoters are active during most embryonic stages and play a direct role in maintaining a reasonable and consistent level of gene expression.

Tissue-specific promoters limit gene expression to certain tissues or involve gene expression in developmental phases. Environmental stimuli have a significant impact on gene expression connected with inducible promoters, allowing for gene control by external causes.

Enhancers are small sections of a gene (50–1500 bp) that can be identified and bound by activator proteins. These proteins, known as transcription factors, bind to the enhancer, generating an enhancer-bound transcription factor complex that subsequently interacts with the mediator complex (TFIID), assisting in the recruitment of RNA polymerase II.

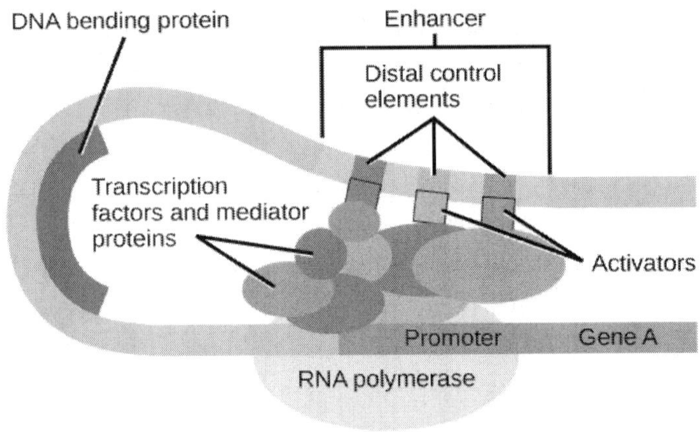

Figure 2.2. A diagram showing Promoters & Enhancers.

Source: Image by Wikimedia Commons

The enhancer-bound transcription factor complex creates a loop and moves toward the intervening sequence, where it comes into contact with the promoter region, enhancing the promoter's accessibility to transcription proteins. Silencers, on the other hand, serve as the polar opposite of enhancers.

Silencers are transcription factors known as repressors that bind to silencers. These repressors are known to inhibit gene transcription. Plant genetic engineering selects appropriate promoters and enhancers depending on the desired control of gene expression.

Whenever the transgenic causes minor toxicity in the target plant, gene expression is kept at a low level, higher gene expression levels, on the other hand, make it easier to detect and monitor a transgenic that is often under-expressed in nature.

2.10.3. Reporter Genes

Reporter genes are the genes that are linked to regulatory sequences or genes of interest in order to monitor transgene expression and the location of produced proteins. A reporter gene sequence encodes proteins or protein products that have been catalyzed for detection using equipment or simple assays.

Selectable marker genes, on the other hand, like antibiotic genes, herbicidal-resistant genesas well as anti-metabolic genes, provide resistance to particular chemical agents that limit nontransgenic plant growth.

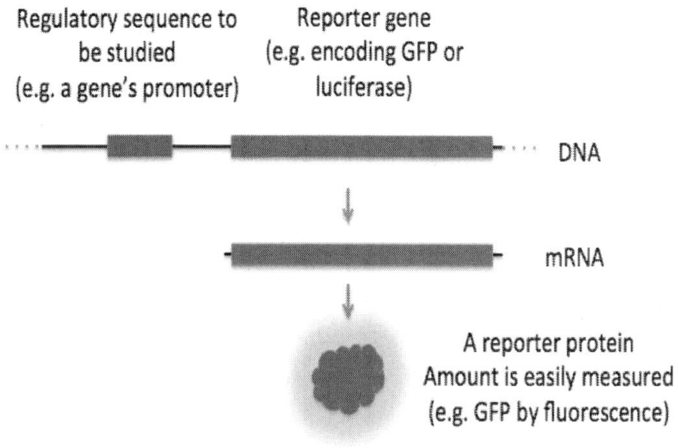

Figure 2.3. A Diagram of Reporter Gene.

Source: Image by Wikipedia

Green fluorescent protein (GFP), chloramphenicol acetyltransferase (CAT), beta-galactosidase (LacZ), luciferase (Luc), & beta-glucuronidase are typical reporter genes used to assess plant transgenic expression (GUS). These reporter genes distinguish transformed from non-transformed cells and allow for the detection of transgene location, and modulation of the expressed & tagged protein.

Dual reporter systems, such as Luc/Luc & GUS/Luc, are also accessible for enhanced protein identification. For large-scale observations throughout a wide variety of cells and tissues, ideal reporter genes must be very sensitive, stable, and dependable.

Despite substantial research into existing reporter systems, the ideal reporter genes with all of the needed features are still unavailable. Each reporter system has both useful and negative characteristics. As a result, the nature of the study must be taken into account while selecting a good reporter gene.

2.10.4. Problem Posed By Antibiotic Resistance Reporter Genes

Plant transformation techniques are now accessible, although they are not flawless. There are no approaches that can guarantee 100 percent transformation efficiency. Markers are required to differentiate between transformed & non-transformed plant cells.

In transforming selection, antibiotic or herbicide resistance genes serve as the principal selective markers, effectively eliminating non-transformants. An antibiotic resistance system's effectiveness is determined by three criteria:

(1) The selective agent employed should totally suppress the development of nontransformed cells,

(2) The resistance gene is amplified in transformed cells, &

(3) The explant utilized for transformation.

Antibiotic screening helped to identify the first successful transgenic plant. Yet, the use of antibiotics constantly raises concerns about environmental issues and the safety of genetically modified (GM) foods. This is mostly due to concerns about gene pollution, which occurs when antibiotic genes escape from GM plants into the environment via microorganisms.

Bacteria are considered capable of absorbing and integrating foreign DNA fragments into their genome. The microorganisms around the GM plant may have absorbed DNA pieces from the transgenic plant and thereby evolved antibiotic resistance.

Furthermore, gene escape is possible because the antibiotic resistance gene can spread to nearby plants via pollen dispersal. Consumption of the transgenic plant with the antibiotic resistance gene can potentially facilitate the transfer of the genes towards the probacteria present in the stomach. As a result, in transgenic whole plant screening, antibiotic resistance marker genes often are avoided.

Efforts were produced by substituting an antibiotic marker gene with another possible marker gene, such as a reporter gene. Reporter genes, such as GFP, are not found to be environmentally hazardous; instead, they are commonly employed as biosensors. Plants engineered with these reporter genes could help to prevent the unnecessary accumulation of antibiotic resistance in the environment.

Figure 2.4. Genetically Modified Trees.

Source: Image by Flickr

Gene escape, on the other hand, may be avoided by removing the antibiotic resistance gene from a transgenic plant. The most recent genome editing techniques, like TALEN and CRISPR/ Cas9, may be useful in removing gene markers from transgenic plants. These sophisticated tools, however, are yet to be developed.

2.11. VECTORS FOR THE PRODUCTION OF TRANSGENIC PLANTS

A vector serves as a carrier for transporting the desired gene into a target cell for replication & expression. A common vector is made up of three parts: a replication origin, a multicloning or recombination site, and a selectable marker.

The source of replication is an AT-rich region on the vector that begins vector replication by attaching to a protein complex, unwinding the vector, and thereby replicating it with the assistance of polymerases. The multi cloning site is an area that comprises numerous distinct sequences, also known as restriction sites, that may be cut by a specific restriction enzyme, enabling the gene of interest to be inserted.

The recombination site facilitates site-specific recombination between two plasmids. The selective markers are thegenetic markers that work as described in the gene construct section, verifying the vector's incorporation into *Agrobacterium* sp. Vectors widely employed in plant transformation are Ti plasmid-based vectors and plant viral-based vectors.

2.12. PLASMID VECTORS

2.12.1. Ti Plasmid

The Ti plasmid is by far the most widely utilized vector in transgenic plant development. The Ti plasmid is expected to be between 200 and 800 kbp in size, depending on the Ti plasmid class. The Ti plasmid is organized into three major regions: transfer DNA (T-DNA), virulence, and opine catabolism.

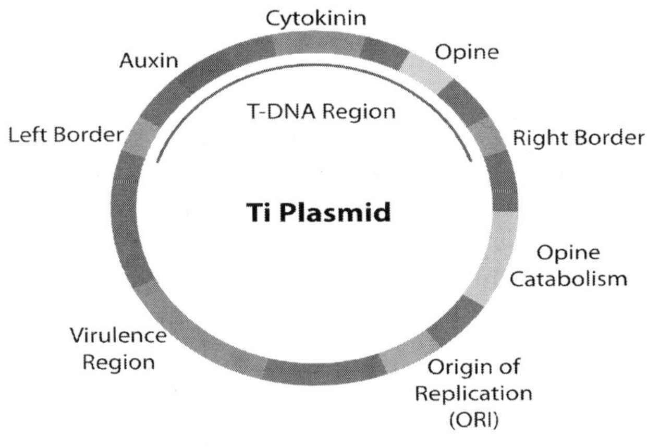

Figure 2.5. Diagram of Ti Plasmid.

Source: Image by Wikimedia Commons

The T-DNA region introduced into the plant genome is approximately 24 kbp in size. This area is bounded on each end by repeat sequences known as the left and right borders. The correct boundary is necessary for the transfer of DNA that causes carcinogenesis. The virulence area, on the other hand, is in charge of encoding the vir genes, which help in the transmission of the T-DNA.

The T-DNA sequence also codes for the manufacture of opine & phytohormones (auxin and cytokinin). The three oncogenes (opine, cytokinin, and auxin biosynthesis gene) found in T-DNA are the primary drivers of tumour growth in plants, resulting in crown gall disease.

Growth hormones produced are responsible for unregulated plant cell proliferation and exacerbate the problem by promoting crown gall development. Opines are the primary carbon source used by *A. tumefaciens* that is not produced spontaneously from plant metabolism.

As a result, by genetically changing the host cells, *A. tumefaciens* will establish its own biosynthetic machinery for nutrition synthesis. The genes for proteins that are involved in opine catabolism are encoded by the opine catabolism portion. The source of DNA replication permits the Ti plasmid to be kept stable in the bacteria.

The Ti plasmid is normally disarmed for plant transformation, with the tumor-inducing genes deleted and replaced with the reporter genes and the gene of interest.

2.12.2. The Co-Integrative Vector

The co-integrative vector is created via homologous recombination of an intermediate vector with a disarmed Ti plasmid. The intermediate vector is often an *E. coli* plasmid containing the desired gene. Both the intermediate vector and the disarmed Ti vector share certain sequences, allowing for homologous recombination of the two plasmids.

The result of the recombination will be a big co-integrative vector, including the merged *E. coli* plasmid and the disarmed Ti plasmid. This co-integrative vector will be returned to the Agrobacterium subsequently for transgenic plant transformation.

However, the massive size of the plasmid as a result of recombination may provide a daunting barrier to manage. As a result, the usage of this vector has been stopped with the introduction of the binary vector system.

2.12.3. The Binary Vector

When researchers discovered that T-DNA could work independently without being attached to the Ti plasmid, they devised a two-plasmid system known as the binary vector system. The binary system included two plasmids, the assistance vector, and the micro vector.

The mini vector is a smaller plasmid that contains both the T-DNA and the origin of replication of *E. coli* and *A. tumefaciens*, allowing the plasmid to be cloned in both *E. coli* and *A. tumefaciens*. The assistant vector is a Ti plasmid that lacks the T-DNA section.

The wild-type Ti plasmid is regarded as a helper plasmid since it contains all of the genes required for gene transfer and integration. Both of these helper and micro vectors are injected into the Agrobacterium at the same time, and also the transformed Agrobacterium is employed in plant transformation.

2.12.4. Plant Virus Vectors

Viruses are obligatory intracellular parasites that replicate via the molecular machinery of a specific host. Viruses have not been discovered infecting plants via transmission vectors like aphids, insects, nematodes, & fungi.

These modified viruses are employed as alternate sources for plant transformation; typical plant viruses used in transgenic plant production include Cauliflower mosaic virus (CaMV), Tobacco mosaic virus (TMV), Alfafa mosaic virus (AMV), Potato virus X (PVX), & Cowpea mosaic virus (CPMV).

In two ways, the wild-type plant viral vectors have been enhanced and modified to allow their employment with Agrobacteria and the plant host for greater efficiency. The first approach would be to create virus vectors that are similar to wild types that carry the desired gene and are capable of infecting plants.

The second strategy would be to create a 'deconstruct' virus, which involves removing unwanted viral genes, such as the coat protein-expressing gene, and replacing them with useful genes, like reporter genes or theantibiotic resistance genes, to allow the transgenic screening.

Figure 2.6. Symptoms of Alfalfa Mosaic Virus on potato leaves.

Source: Image by Wikimedia Commons

2.13. TRANSFORMATION TECHNIQUES FOR THE PRODUCTION OF TRANSGENIC PLANTS

Plant transformation is the process of changing the genetic contents of an interesting plant by inserting DNA segments into the plant genome to reach optimal gene expression. A wide range of plant transformation techniques are now available to the general population. These plant transformation approaches are classified as either indirect or direct gene transfer.

Indirect gene transfer (also known as vector-mediated gene transfer) involves the insertion of exogenous DNA into the plant genome using biological vectors, while direct gene transfer methods include inserting exogenous DNA directly into the plant genome by physical or chemical interactions.

2.14. IN-DIRECT METHODS

2.14.1. Agrobacterium-mediated Gene Transfer

Because it is efficient and effective in a wide range of plants, Agrobacterium-mediated transformation is by far the most often employed approach in plant transformation. Agrobacteria are naturally occurring in the soil ecology.

They are Gram-negative pathogenic bacteria that cause crown gall or hairy root infection in plants. The genetic information for tumor development is contained in the genome of these bacteria on a tumor-inducing plasmid (Ti plasmid) or a hairy root-inducing plasmid (Ri plasmid).

Agrobacterium tumefaciens&*Agrobacterium rhizogenes* are the two most frequent Agrobacterium species employed in plant transformation. The Ti plasmid produces crown gall disease in *A. tumefaciens*, whereas the Ri plasmid causes hairy root disease in *A. rhizogenes*.

When the harmful genes in Agrobacteria are eliminated, the discovery of the two species provides efficient vector systems for the growth of transgenic plants. This approach effectively converted a wide range of plants, including rice, maize, barley, & tobacco.

The indirect approach of genetic transformation uses bacteria as a vector to transfer the gene construct into a target cell. For plant transformation studies, *Agrobacterium* (a gram-negative soil bacteria that causes crown gall disease in many plants) is used.

Agrobacterium tumefaciens, the best-researched species in this genus, create an effective delivery mechanism for genetic transformation in plants. These bacteria are infected with a big plasmid termed Ti plasmid (tumor-inducing), which contains tumor-inducing genes (T-DNA) and certain other genes required in T-DNA integration into the host genome.

Wounded plants release sap rich in phenolic chemicals, which function as chemical attractants for the Agrobacteria and drive vir gene production. It results in Agrobacterium infection of the plant and the insertion of a T-DNA region at a random spot in the host genome, and plant cell proliferation to create crown gall expansion.

A. tumefaciens, which are employed for plant transformation, is modified a Agrobacteria with no tumor-promoting or opine-synthesis genes in the genome. These genes are extracted (disarmed) from bacterial plasmid & replaced with the necessary foreign gene or selection markers, making them excellent vectors for temporary or stably incorporating foreign genes into plant genomes.

To achieve stable gene incorporation, Agrobacteria transport and integrate T-DNA into the host's genome via the following steps:

(1) Recognition of the host's chemical signal,

(2) Initiation of the vir gene in Agrobacterium,

(3) Attachment of Agrobacterium to plant cells,

(4) Activation & transit of virulence proteins,

(5) Production of T-DNA strand,

(6) transfer of T-DNA & virulence protein out of Agrobacterium,

(7) T-DNA translocation into plant nuclear,

(8) T-DNA integration into plant genome excluding steps

(7) The stages involved in temporary transformation are thought to be similar to those involved in the stable transformation

Because the Agrobacterium-mediated transformation methodology was first presented, several refinements have been made to increase its efficiency. Although classic Agrobacterium-mediated transformation works well in dicotyledonous plants like potatoes, tomatoes, and tobacco, it is less effective in resistant crops like wheat and maize owing to a lack of a wounding response system.

This remained a significant barrier until the invention of plant tissue culture as well as the launch of in-planta transformation protocols, which greatly enhanced the transformation efficiency of these plants.

Agrobacterium-mediated in-planta transformation does not use plant tissue culture; instead, transformation is performed directly on a grown plant. This method incorporates agro-inoculation and agro-infiltration.

The modified Agrobacteria with the gene of interest is implanted into the surface of the plant tissue of a complete plant in agro-inoculation. It is usually done using a toothpick, a wire loop, or direct organ absorption (loral dip method).

Agroinfiltration, on the other hand, can be done with a syringe or a vacuum. Syringe infiltration is easy and inexpensive since it injects the converted Agrobacteria into the underside of the leaf while simultaneously applying counter force on the other side of the leaf. However, it is time-consuming and only suitable for small-scale expression.

However, vacuum infiltration is faster and more efficient, allowing for large-scale production and commercialization. The entire plant is submerged in the converted Agrobacteria suspension, and a vacuum atmosphere is used to push the Agrobacteria to permeate throughout the entire plant. These alternate techniques of plant transformation have gained popularity, particularly in monocots.

Heat & hydrolysis treatment on target tissues before transformation have been observed to improve transformation efficiency when heat treatment is employed, improving transformation efficiency in different plant tissues such as switchgrass, ryegrass, and rice.

A similar improvement was found in hydrolysis treatment using hydrolytic enzymes like cellulase, macerase, and pectinase, which offer softer disruption and improves the recovery and regeneration rate of changed cells.

Furthermore, sonication-assisted plant transformation, which was administered to the target plant tissues before Agrobacteria immersion or agroinfiltration, resulted in an effective transformation of refractory plants by producing micro-wounds on explants, which allows for better access to Agrobacterium.

2.15. DIRECT METHODS

2.15.1. Direct Gene Transfers

As the name implies, direct gene transfer includes the direct introduction of external DNA (naked DNA) into the plant nucleus. To introduce foreign DNA into a plant cell, the cell's outer membrane is first broken, allowing foreign DNA to enter.

The majority of direct gene transfer procedures are straightforward and effective. Yet, gene expression in these transgenic plants can be converted transiently or persistently.

Direct gene transfer is classified into two types: physical gene transfer & chemical gene transfer. Physical gene transfer causes mechanical disruption of the cell wall and membrane. Since it was first developed by Sanford et al., particle bombardment biolistic has been the most often utilized approach in plant transformation.

Using a "Gene Gun" (Helios® Bio-Rad), DNA covered with gold or tungsten particles is fired into the target plant cell at high pressure. The fast-moving particles enable coated DNA to penetrate the thick plant cell wall, guiding the foreign DNA into the plant's nucleus. The coated DNA will subsequently detach from the metal particles and merge into the plant cell's chromosomes within the nucleus.

This approach was discovered to be effective in converting both dicots and monocots, making up for the less successful Agrobacterium-mediated transformation process. It is also less harmful and may be used on practically all plant cells. The significant drawbacks of this procedure, however, are the lack of specific tools and the inability to transfer DNA fragments to the plant nucleus rather than other organelles.

Microprojectiles (gold or tungsten) are generally covered with DNA in the presence of calcium chloride & spermidine in the classic biolistic approach. Spermidine helps to maintain the DNA structure and improves DNA binding to microprojectiles.

Other cationic polyamines, such as protamine, give better outcomes when compared to spermidine in the effort to improve this technique since this assures by preserving the coated DNA from DNase destruction. Biolistic transformation using protamine was done in bothrice and peanut, and the outcomes were thrice better than spermidine.

Other approaches to enhancing transformation efficiency using biolistic guns included reducing the quantity of DNA covered on the microcarriers.

Electroporation, which employs electrical impulses to assist the transfer of foreign DNA into plant cells, is another physical gene transfer method. Plant cells are first incubated in a buffer solution that contains foreign DNA, and then electrical impulses are applied to the buffer, resulting in the creation of brief transient holes in the plant's cell membrane, allowing the foreign DNA to penetrate.

This approach is simple and quick, but it only works with protoplasts (cells without cell walls). As a result, this approach is rarely used in plant transformation.

Chemical gene transfer methods involve the employment of chemicals to damage cell membranes, allowing foreign DNA to enter. This approach is not preferred for plant transformation since it is only effective when administered to protoplasts.

One of the most popular compounds utilized in this strategy is polyethylene glycol (PEG), which is used to destabilize the cell membrane in the presence of a divalent cation, enhancing cell membrane permeability and allowing for the absorption of foreign DNA.

The precise process for chemical gene transfer is unknown; however, it has been proposed that PEG raises osmotic pressure and promotes protoplast contraction, facilitating endocytosis of the divalent cation/DNA complex. Aside from them, liposome is a chemical technique employed in the transformation of plant protoplast cells.

Liposomes serve as transport vehicles for foreign genetic elements entering the protoplast. Liposomes' lipophilic properties allow for simple entry into the protoplast during cell transformation.

2.15.2. Microprojectile/Particle Bombardment (Biolistics)

Biolistics is a technique that involves physically impregnating cells with nucleic acids or other biological materials. The biolistic particle delivery system is a plant transformation tool that bombards cells with heavy metal particles coated with DNA/RNA.

John Stanford created this approach in 1984 for physically introducing DNA into cells in order to circumvent Agrobacterium host-range limits. The Agrobacterium-mediated genetic transformation technique is effective for dicotyledonous plants but inefficient for monocots.

The biolistic particle delivery system is an efficient and adaptable method for transforming practically any type of cell. It has been shown to be a viable method for producing transgenic organisms in prokaryotes, animal, and plant species.

In this procedure, a construct containing the desired gene is coated on the surface of small gold or tungsten particles (0.6 – 1 mm in size). DNA is precipitated using calcium chloride, spermidine, and polyethylene glycol prior to coating. These coated microparticles are placed onto the macro-carrier and rapidly accelerated using compressed helium gas.

These microprojectiles might be directed towards plant cell suspensions, callus cultures, or tissues. As the microprojectiles pass through the plant cell walls and membranes, coated DNA is liberated from their surface and integrated into the plant's genome. The usage of binary vectors containing T-DNA border sequences is not essential in biolistics.

2.15.3. Electroporation

Electroporation is a method of direct gene transfer transformation. In this procedure, a cell and DNA combination is subjected to very high voltage electrical pulses (4000 – 8000 V/cm) for very brief periods of time (a few milliseconds). It causes transitory holes to develop in the plasma membrane, via which DNA seems to enter the cell and ultimately the nucleus.

A cell suspension containing plasmid DNA is inserted in an electroporation cuvette between electrodes, and electrical pulses are administered. Cell membranes create temporary micropores that allow cells to take in plasmid DNA, resulting in steady or transitory DNA expression. Cells halted during the metaphase stage of the cell cycle are highly favorable for electroporation because they lack a nuclear envelope and exhibit unique plasma membrane permeability. Because strong plant cell walls hinder DNA transport, protoplasts are employed for electroporation of plant cells.

The electroporation method was initially created for protoplasts, but it has also shown excellent results with cells and tissues, allowing for rapid restoration of regenerated plantlets. Electroporation has also been used to create transgenic maize using immature zygotic embryos & embryogenic calli.

When compared to organized tissues, protoplast transformation is linked with low transient expression of transgenes and low regeneration frequency, particularly in monocotyledonous plants. The electrical field & chemical compounds used to disorganize cell walls significantly impair protoplast viability and division capabilities.

2.15.4. Microinjection

Microinjection is the method of manually injecting transgene at a microscopic or borderline macroscopic level using a tiny glass micropipette. The transgene might be circular or linear and can be delivered in the form of plasmids, cosmids, phage, YACs, or PCR products.

Using a glass microcapillary injection pipette, DNA is directly mechanically introduced into the nucleus or cytoplasm. The protoplasts are trapped in low melting agar using a holding pipette & suction force under a microscope.

The DNA is then delivered straight into the cytoplasm or nucleus. *In-vitro* cultivation of the implanted cells is followed by plant regeneration. This method has been demonstrated successfully in rapeseed, tobacco, and a variety of other plants.

This approach can produce stable transformants, but it takes technical skill and is a time-consuming procedure. Furthermore, due to the thick cell walls of plants and the lack of a single cell-to-plant regeneration pathway in most plant species, Microinjection has only met with little success in plant transformation.

A regular compound microscope (about 200X magnification), an inverted microscope (approximately 200X magnification), or a dissecting stereomicroscope (around 40-50x) are employed in this approach. The target cell is placed under the microscope, and the cell membrane & nuclear envelope are pierced using two micromanipulators.

One micromanipulator is holding the pipette, while the other is holding the microcapillary needle. Microinjection systems are classified into two types: continuous flow systems and pulsed flow systems. The amount of the sample injected in the continuous flow system is dictated by the length of time the needle spends in the cell.

The continuous flow system is simple and economical, but it is out of date. The pulsed flow technology allows for better control over the volume of material given, needle location, and movement. Although this technology causes less harm to the receiving cell, the components of this system are relatively pricey.

2.15.5. Chemical Mediated Gene Transfer

Some substances can be used to induce cells or protoplasts to take up foreign DNA. The most used chemical for this function is polyethylene glycol (PEG). It aids in the precipitation of DNA, which is subsequently taken up by the calls via the endocytosis process.

2.15.6. Liposome Mediated Gene Transfer

Plasmids encoding desirable foreign genes can be wrapped in little lipid bags known as liposomes, which can subsequently be fused with protoplasts by using chemicals such as PEG.

2.16. CONCLUSION

The sequences of reporter genes encode proteins or associated proteins that have been processed for detection using apparatus or simple tests. Despite extensive study into current reporter systems, optimal reporter genes with all of the required properties remain elusive.

As a result, while choosing an appropriate reporter gene, the purpose of the investigation must be considered. There are no methods that can ensure complete transformation efficiency. Antibiotic or herbicide resistance genes serve as primarily selected markers in transformant selection, effectively removing non-transformants.

However, the use of antibiotics is continually raising worries about the environment and the safety of genetically modified (GM) foods. The T-DNA sequence also encodes for the synthesis of opine and phytohormones (auxin and cytokinin).

The binary system had two plasmids, an aid vector, and a micro vector. The mini vector is a smaller plasmid that includes both T-DNA and E's replication origin. Viruses are intracellular parasites that multiply via the molecular machinery of their host.

REFERENCES

1. Aghaee, K., Akhshi, N. and Zebarjadi, A., 2014. *Application of transgenic plants as factories for producing biopharmaceutical*. [ebook] Available at: <https://www.researchgate.net/publication/268514150_Application_of_transgenic_plants_as_factories_for_producing_biopharmaceutical> [Accessed 7 June 2022].
2. Berkowitz, G., 2016. *The Future of GMO Crops | Science of GMOs*. [online] Gmo.uconn.edu. Available at: <https://gmo.uconn.edu/topics/the-future-of-gmo-crops/#> [Accessed 7 June 2022].
3. Gktoday.in. 2018. *Advantages and Disadvantages of Genetically Modified Crops - GKToday*. [online] Available at: <https://www.gktoday.in/topic/advantages-and-disadvantages-of-genetically-modified-crops/> [Accessed 7 June 2022].
4. Low, L., Yang, S., Kok, D., Ong-Abdullah, J., Tan, N. and Lai, K., 2018. Transgenic Plants: Gene Constructs, Vector and Transformation Method. *New Visions in Plant Science*, [online] Available at: <https://www.intechopen.com/chapters/63134> [Accessed 7 June 2022].
5. Microscopemaster.com. n.d. *Transgenic Plants*. [online] Available at: <https://www.microscopemaster.com/transgenic-plants.html> [Accessed 7 June 2022].
6. n.d. *Methods of Gene Transfer*. [ebook] Institute of Lifelong Learning, University of Delhi. Available at: <https://www.deshbandhucollege.ac.in/pdf/resources/1589512616_Z(H)-VI-Biotech-1.pdf> [Accessed 7 June 2022].
7. Thakshin, A., 2020. *Transgenic plants Advantages & Disadvantages*. [online] TheQnA.Org. Available at: <https://theqna.org/transgenic-plants-advantages-disadvantages/> [Accessed 7 June 2022].
8. Zerbini, F., Silva, F., Urquiza, G. and Basso, M., 2014. Transgenic Plants. *Biotechnology and Plant Breeding*, [online] pp.179-199. Available at: <https://www.sciencedirect.com/science/article/pii/B9780124186729000088?via%3Dihub> [Accessed 7 June 2022].

CHAPTER 3

PLANT METABOLOMICS: AN EMERGING TECHNOLOGY FOR CROP IMPROVEMENT

CONTENTS

of overlap in the way these phrases are used in the field of human disease research, and they are frequently used interchangeably.

In conjunction with the rising ability of bioinformatics, techniques used to evaluate metabolites have exhibited extraordinary refinements in recent decades, such as advancements in mass spectrometry (MS) and nuclear magnetic resonance spectroscopy (NMR). Metabolomics can be used for functional genomics in crops, as well as how it can be combined with crop breeding to produce future crops.

The domains of genomics, transcriptomics, epigenomics, proteomics, metabolomics, and phenomics have all seen significant advancements in recent years. These 'Omics' approaches have improved the precision and speed of ongoing breeding projects in generating climate-smart and nutrition-rich germplasm, which is critical for ensuring food security.

In recent years, the importance of phenomics-based breeding in enhancing agricultural performance has been recognized, and genomics has also played a significant role in generating larger genetic gains. Nonetheless, the various omics platforms show a lot of promise in terms of increasing our existing understanding of critical features and allowing us to build novel plant improvement tactics.

The most sophisticated of the omics techniques is metabolomics, which has received insufficient attention in crop science, particularly for trait mapping and plant selections.

An indispensable component of plant metabolism, the metabolites owe it to their influence on plant biomass and architecture. Metabolomics has established itself as a major breakthrough in science in recent years, paving the way for precise metabolite profiling in microbes, plants, and animals.

It can identify a wide range of metabolites from a single extract, allowing for quick and accurate metabolite analysis. To put it another way, metabolomics provides a comprehensive perspective of cellular metabolites such as tiny organic substances that participate in many cellular activities, thus portraying the cell's absolute physiological state.

In light of the fast-growing field of metabolomics, metabolite analysis of mutants and transgenic lines has the potential to help researchers better understand metabolic networks and identify the underlying candidate gene(s).

Furthermore, metabolomics aids in the resolution of gene function: how a specific gene affects the metabolic route, as well as uncovering numerous layers of regulation and interception between linked pathways, which is hard to achieve using traditional assays such as microarray.

Researchers will be able to classify and prioritize genes to improve essential features in crop species using an integrated method that incorporates inferences from genomes, transcriptomics, proteomics, and metabolomics.

Furthermore, the above-mentioned omics investigations have been expanded to include regulatory stages such as epigenetic regulation, post-transcriptional modification, and post-translation modification.

For that purpose, interactome network research aimed at uncovering molecular interactions between biomolecules (nucleic acid, proteins, amino acids, carbohydrates, lipids, and so on) add to our understanding of the genotype-phenotype relationship.

Regardless of the availability of transgenic systems, metabolomics is becoming increasingly popular in many crop species. Metabolomics has the potential to make superior trait selection and breeding material improvement easier.

The availability of whole-genome sequences, genome-wide genetic variations, and cost-effective genotyping assays, together with improvements in metabolomics, creates an exciting potential to efficiently integrate metabolomics into crop breeding programs.

The methods and tools used in the study of metabolomics, including mass spectrometry (MS) and nuclear magnetic resonance (NMR) spectroscopy, have witnessed significant betterment. The presently available metabolomics platforms have the capacity to enable large-scale metabolite surveys covering both known as well as unknown metabolites.

The deluge of such data, however, makes the annotation of metabolites a considerable challenge. In this context, the ever-growing strength of bioinformatics tools coupled with the establishment of metabolomics databases such as the one for the model plant *Arabidopsis* and others for various plant species have greater implications for metabolite annotation.

A considerable amount of data have resulted from metabolic surveys, which might support plant improvement schemes focusing on the traits of agricultural importance such as yield and stress tolerance.

Further, rapid generation of genome-scale data by sequencing of DNA and RNA and by MS quantification of proteins and metabolites necessitates the integration of this information in order to devise a holistic way of improving traits of interest.

Although the majority of current research is conducted on well-established model organisms, similar research might be conducted on other plant species as well. Dealing with enormous multi-omics data for undertaking systems-level analysis is now a daunting task for the scientific community.

Improved statistical and bioinformatics techniques will be necessary for this scenario to examine these data sets together for better consolidation, which can then be translated to better plant performance. In this section, we briefly discuss the most recent research on plant metabolites and the use of metabolomics, including metabolic engineering, to improve plants.

3.2. DIFFERENT PLATFORMS TO GATHER METABOLOMIC DATA

Take the example of tomato as a model system that contains distinct categories of chemical compounds contributing to the fruit quality, including sugars, organic acids, amino acids, fatty acids, isoprenoids, and polyphenolic compounds.

A variety of separation methods has been used to investigate the tomato metabolome, using both targeted and nontargeted metabolomics, leading to a wide range of quality biomarkers.

Because most research projects focus on understanding or improving a particular target feature, targeted metabolomics is by far the most prevalent method. There is a wealth of information available to explain the phenotypic variance, but it may not be readily available.

Small molecules can have great effects. For example, the variation in the ratio between sweetness and acidity results in tomatoes having a sharp, sweet, insipid, or lovely taste. Accelerating improvements through breeding programs require large-scale and low-cost assays that enable the analysis of thousands of samples within a short period of time.

Phenotypic surveys of diverse germplasm have a very broad scope and help define the range of acceptable phenotypic variation, albeit limited in their depth. Such kinds of data on organic acid and sugar can be used in conjunction with gene expression analysis to determine the genetic origins of poor fruit quality.

The information on carbohydrates and organic acids can also be obtained by employing more sophisticated tools such as nuclear magnetic resonance (NMR) spectroscopy, which detects more compounds per assay than enzymatic or colorimetric methods but at a much lower throughput.

NMR spectroscopy is used for the structural determination of a novel metabolite of particular interest. Alternatively, gas chromatography (GC) paired with mass spectrometry (MS) (GC–MS) permits broad-scope metabolomic profiling, with increased throughput in comparison to the NMR.

On the other hand, the requirement for chemical derivatization by GC–MS may result in the exclusion of some metabolites from the study, as well as a lack of sufficient information for a definite identification of a specific metabolite. Combining numerous datasets from complementing analytical platforms, on the other hand, is a powerful technique for analyzing metabolomes.

In tomatoes, color and aroma are other targets for improvement. A majority of pigments in tomatoes are isoprenoids, such as carotenoids, while others are polyphenolics (e.g., flavonoids). Traditionally, liquid chromatography (LC) with commercial standards is used for carotenoid profiling. However, LC-MS is to be used for a complete estimate of metabolomes, especially for isoprenoids.

The MS analysis can be done in real-time with the LC or offline. Inline MS streamlines workflow; however, offline MS may improve sensitivity by reducing sample complexity. NMR spectroscopy could also be used for isoprenoid profiling, which can distinguish between E and Z isomers and is not possible with MS analysis.

This is significant because different carotenoid isomers may have varied biological activity and, as a result, nutritional properties. Carotenoid composition changes throughout food preparation and processing, both in terms of quality (i.e., isomerization) and identity (i.e., oxidation) (i.e., degradation by heat).

For a thorough description of the isoprenoids, an examination of both raw and cooked samples is required. In addition to color, carotenoids also contribute to fruit aroma, as do fatty acid and amino acid derivatives. All three represent volatile compounds, GC and GC–MS are used for their separation and identification. LC-MS and MS/MS were used to conduct a metabolite survey of about 100 Dutch tomato varieties.

One of the common issues encountered while examining MS or NMR data in order to better comprehend the spectra produced during an experiment is the need for a well-curated database. Thankfully, recent advances in tomato metabolomics have resulted in the emergence of such community-oriented tools.

Multiple software and analysis tools have been created in recent years to process and evaluate metabolite data, but none of them is self-sufficient enough to meet user expectations.

In this regard, the Government of India's Department of Biotechnology has launched a project to establish a web-based collaborative platform for metabolomics researchers to store, analyze, and share their data (Computational Core for Plant Metabolomics, CCPM).

3.3. GENE IDENTIFICATION

Metabolomics study assists in identifying particular mQTL, which corresponds to gene(s) related to that particular trait. The approach is increasingly gaining recognition because once mQTL is identified, then it becomes easier to single out the gene(s) rcsponsible for that certain metabolite.

The essential aspect of bioinformatics is gene prediction. It involves locating regions of genomic DNA that encode genes (protein-coding genes). Gene prediction or gene identification is highly important owing to the help it provides to the scientists to differentiate between coding and non-coding regions of a genome, functionally explain genes, conduct research related to detection and treatment, prevent genetic disorder diseases, etc.

3.4. METABOLOMICS IS AN EMERGING AND EVOLVING TECHNOLOGY

Metabolomics presents a significant analytical challenge in practice because, unlike genomic and proteomic methods, it aims to measure molecules that have disparate physical properties (e.g., ranging in polarity from very water-soluble organic acids to very nonpolar lipids.

As a result, most comprehensive metabolomic technology platforms divide the metabolome into subsets of metabolites—often based on compound polarity, similar functional groups, or structural similarity—and design sample preparation and analytical techniques that are optimized for each.

The metabolome is hence, measured as a patchwork of results from different analytical methods. The methodologies employed in metabolomics continue to progress and improve as an emerging area that has been facilitated, at least in part, by the continual development of analytical instrumentation with new capabilities each year.

Individual laboratories tend to have distinct methodologies, and there are very few standard operating procedures that are generally followed across laboratories as a result of metabolomics laboratories using various procedures that are potentially subject to regular improvement.

Although this diversity of technology is linked to field innovation, it can provide problems when comparing data between laboratories due to concerns such as measurement precision for certain classes of metabolites or nonoverlapping metabolite coverage.

Additionally, the degree of certainty in metabolite identification can vary among methods, ranging from metabolite identities rigorously confirmed using authentic standards of reference to putative identifications made employing reference databases to signals that remain as "unknowns."

The need for standardization in metabolomics has been lauded by its practitioners, giving rise to various initiatives toward realizing this aim, such as the Metabolomics Standards Initiative to develop guidelines for data reporting ring tests to evaluate the abilities of different metabolomics methods and laboratories to obtain similar results.

Open-access repositories for metabolomics results and related metadata, such as the Metabolites database in Europe and the Metabolomics Workbench funded by the National Institutes of Health Common Fund in the United States.

With time, as the field matures, it is conceivable that convergence among laboratories on particular standard operating procedures might occur; however, in the meantime, methods must be evaluated along with the generated results on a case-by-case basis.

Comprehensive descriptions of methods, including details such as instrument settings and the means by which each metabolite was identified, are required information that can aid this evaluation; data from repeated measures of representative samples, such as pooled samples, to enable the determination of coefficients of variation for each metabolite measured by a particular method.

Access to data sets submitted either as electronic supplements or to metabolomics data repositories is also required information that can aid this evaluation.

3.5. PLANT METABOLOMICS USED FOR CROP IMPROVEMENT

3.5.1. Metabolomics for Improvement of Rice

Worldwide, rice is an essential staple crop. It has benefitted considerably from the developments in the field of genomics. For example, the rice genome has been sequenced and is found to encode approximately 32,000 genes. More than half of these genes, however, have yet to be identified as to what they do biologically.

Using gain-of-function and loss-of-function techniques, new genes in rice have been discovered. Genetic linkage and association analyses with genetic core collections and segregating populations have been utilized to examine the direct relationships between metabolic composition, genotypes, and phenotypes as representatives for agronomical traits.

Since rice is one of the most important crops in the world, there is a constant challenge to generate better rice cultivars with valuable traits, although it has been performed using traditional and molecular breeding techniques.

Recently, it has been found that metabolomic methods help to generate new rice cultivars in the context of identification of valuable gene functions and safety assessment of GM crops.

In this section, we elucidate actual examples of applications of metabolomic techniques to rice studies from two viewpoints: the application of metabolomics to the study of rice biology, including the detection of metabolic changes according to the natural genetic variations in rice plants, and the application of metabolomics to the production of GM rice, including the evaluation of substantial equivalence.

Natural variation analysis in rice using metabolomic techniques is considered useful not only for understanding biological traits of rice, such as yield and defense responses but also for improving rice quality, including taste and nutritive value.

Kusano et al. used a metabolomic method combining one-dimensional (1D) and two-dimensional (2D) GC-TOF MS for metabolic phenotyping of the 68 rice varieties in the world rice core collection (WRC) and two other varieties. The WRC is a representative set of the cultivated rice (*Oriza sativa* L.) and covers approximately 90% of the DNA polymorphisms detected in the original population.

In order to characterize the metabolic phenotypes of WRC, metabolomics analysis of the extracts of brown rice seeds was conducted as follows. First, 1D-GC-TOF MS analysis was combined with multivariate statistical analysis, i.e., PLS-DA, to narrow down the representative varieties, which were chosen based on the PLS-DA criterion of good separation from the other varieties.

Second, for high-resolution metabolite profiling, the three selected varieties were subjected to a detailed 2D-GC-TOF/MS analysis. Finally, researchers statistically compared the ten metabolites whose concentrations differed between the three selected cultivars and a control rice cultivar (*Nipponbare*).

The selected varieties were successfully defined in terms of differences in their metabolite compositions as a result of these experiments. For instance, although the selected variety *Urasan 1* is considered to be a *Japonica* variety on the basis of its DNA polymorphisms, this variety showed a unique metabolic profile when compared to other *Japonica* varieties.

This finding suggests that the metabolomic approach can be used to assess the quality of rice cultivars based on their metabolite composition.

Using metabolomic techniques, changes in metabolite concentrations were investigated according to developmental or diurnal periods. The combination of GC–MS and statistical analyses were employed to discover certain biomarker metabolites that occur in rice plants during the developmental period.

Rice samples in different developmental stages were extracted through the use of chloroform and water, and the hydrophobic and hydrophilic extracts were studied by GC–MS. Next, the metabolome data was projected onto PCA.

The fundamental component representing the differences in the developmental period of rice was determined, and then the loading metabolites on this axis were found to be candidates for the biomarker metabolites related to the developmental period of rice. Finally, the physiological roles of each biomarker candidate were discussed.

This sequential flow from sample extraction to statistical analysis appears to be the norm for current metabolomic studies. Moreover, Sato et al. applied CE–MS and CE–DAD (DAD, diode array detector) technology to detect water-soluble metabolites in rice leaves during the day and night.

The levels of 88 metabolites were measured, including amino acids, organic acids, nucleotides, and sugars, whose detection limits ranged from 0.3 to 71 μmol·l−1. Metabolite fluctuations within a single day were discussed further by comparison with the results reported in the earlier investigations.

Additionally, there have been reports on the extensive metabolite analysis of rice plant root exudates and flavor volatiles. Thus, there have recently been numerous examples of metabolomic techniques being applied to rice plants, indicating that metabolomics is recognized as a powerful tool for studying rice biology.

3.6. APPROACHES TO COLLATE METABOLITE, PHENOTYPIC, AND GENOTYPIC DATA: SOME EXAMPLES IN RICE ARE AS FOLLOWS:

3.6.1. Gain-Of-Function Approach

The FOX hunting system enabled the development of the rice full-length (FL) cDNA collection (*Oryza sativa L. ssp. japonica "Nipponbare"*). The FOX hunting system is unique in that it allows for ectopic expression of any plant FL-cDNA library, even in heterologous plant systems, allowing for functional gene analysis.

More than 30,000 transgenic *Arabidopsis* lines expressing rice FL-cDNAs have been created, dubbed "rice FOX Arabidopsis lines." With these FOX lines, metabolic fingerprinting and metabolic profiling were used to identify functional genes in rice.

A nondestructive analytical method based on Fourier transform-near-infrared (FT-NIR) spectroscopy was developed to screen a large number of rice FOX *Arabidopsis* lines. Unlike MS techniques, FT-NIR analysis does not require destructive preparation and can collect data in as little as one minute.

The authors used FT-NIR to analyze approximately 3000 FOX seeds to obtain their metabolite fingerprints. The analysis of the re-transformants' metabolite fingerprint changes resulted in the identification of seven lines with altered metabolite fingerprints in seeds.

Annotations for inserted FL-cDNAs can be found in five of these seven lines. The association of the genes with biological processes emphasized the importance of complex networks underlying plant metabolomic responses.

3.6.2. Loss-Of-Function Approach

TheTos17retrotransposon- andDs-transposon-inserted mutant lines have served as loss-of-function resources for the characterization of the novel genes in rice.Tos17-knockout lines characterize glutamine synthetase (GS) and catalyze the key step of ammonium assimilation.

Tabuchi et al. (2005) used the Tos17-retrotransposon inserted lines to show that the three genes (OsGS1;1, OsGS1;2, and OsGS1;3) encoding cytosolic GS (GS1) in rice. The OsGS1;1 gene was critical for normal growth and grain filling. They used a GC-TOF-MS-based assay to investigate the mutants' metabolomic changes and metabolite-to-metabolite correlations.

In comparison to wild-type rice, the mutants had significantly higher levels of sugars and sugar phosphates while having lower levels of amino acids and rice leaf TCA cycle intermediates. In the presence of ammonium, changes in metabolite profiles differed between root and leaf parts. Surprisingly, nitrogen-containing secondary metabolites were found in abundance.

The study discovered new relationships between the mutant roots' over-accumulated metabolites and some primary metabolites. These findings show that OsGS1; 1 plays an important role in regulating the global metabolic network in rice plants grown with ammonium as the nitrogen source.

3.6.3. Metabolomics for GM Rice

To improve the nutritional quality of GM rice, molecular breeding has been attempted. One of the manipulation targets is the engineering of a metabolic pathway to produce rice cultivars with a high content of valuable metabolites, such as β-carotene in "golden rice" and tryptophan.

Figure 3.2. The Golden Rice plants in the IRRI greenhouse produce these yellow grains.

Source: Image by Flickr

Because tryptophan is an essential amino acid for mammals and a growth-promoting fodder for livestock, high tryptophan rice is a potential GM crop. Because tryptophan can now be produced industrially through fermentation, high tryptophan GM rice could be used as a functional food or fodder with a health or growth-promoting benefit.

Plants normally maintain a low level of tryptophan due to tryptophan's feedback inhibition of anthranilate synthase (AS); hence, AS has been considered to be the prime target in order to increase the amount of tryptophan.

Tozawa et al. cloned 2 AS α-subunit genes (OASA1, OASA2) from rice (*Oryza sativa cv. Nipponbare*) and produced transgenic rice expressing a feedback-insensitive OASA1 gene (OASA1D) in which aspartate-323 is replaced by asparagine.

Detailed evaluation of OASA1D transgenic rice revealed that the levels of free tryptophan in the calli, seedlings, and seeds were increased up to 35- to 300-folds, while the other agronomic traits in the field-grown transgenics were essentially identical to those of untransformed variety.

In addition to the visible phenotypes, it is possible that recombination of tryptophan biosynthetic genes affected other invisible phenotypes, such as the concentration of metabolites connected or unconnected to the tryptophan biosynthetic pathway. Indeed, a change in the pool size of tryptophan in potatoes has been shown to activate the biosynthesis of phenylpropanoid compounds.

Also, because tryptophan is a precursor for the biosynthesis of many plant metabolites, including auxin, indole-3-acetic acid, and indole alkaloids, some of which are toxic or bitter to mammals, tryptophan overabundance is likely to be linked to the production of these by-products.

However, predicting the side effects has been difficult due to our limited understanding of plant metabolism regulation. Thus, it is critical to use metabolomic techniques as well as other omics tools to survey a wide range of metabolites in order to assess the substantial equivalence of GM crops.

Microarray and targeted metabolite analyses of OASA1D rice seedlings revealed that the transgene had a very limited effect, most likely due to the low activity of tryptophan-utilizing pathways in rice seedlings. Profiling analyses of UV-active metabolites in rice calli and seeds expressing the OASA1D transgene revealed that the drastic accumulation of tryptophan had little effect on the profiles of other UV-active metabolites in these tissues, with the exception of the accumulation of a few minor indole alkaloids.

According to these findings, the unchanged metabolic profiles are the result of low secondary metabolic activity in rice, implying that transformation with OASA1D could be useful in breeding crops with higher tryptophan content.

In addition to the above-mentioned studies, there have been a number of reports regarding the application of metabolomics to the production of GM rice.

For example, to understand the correlation between gene overexpression and the concentrations of metabolites, GM rice overexpresses theYK1gene, which is known as the homologue of the HC-toxin reductase gene and has a dihydroflavonol-4-reductase activity, was subjected to metabolomic analysis using FT-ICR MS and CE–MS.

The number of amino acids and organic acids in the leaves, roots, seeds and calli of this GM rice changed slightly. Simultaneously, an increase in the level of glutathione derivatives was discovered in the calli. This report summarizes the findings regarding the relationship between YK1 gene overexpression and glutathione derivative accumulation.

Furthermore, several types of low phytic acid-containing mutants (LPA) were analyzed using GC for the elucidation of the varieties of mutants that have a small amount of phytic acid considered to be an antinutrient factor in rice grains. The metabolic changes in mutant rice grown in various fields were also studied. The effects of the fields in which the rice varieties were grown and the mutations on the concentrations of various metabolites were discussed using these metabolomic data. As these studies show, metabolomic approaches may help us understand the effects of genetic modifications on plant metabolite concentrations.

Molecular breeding of GM rice through the introduction of various transgenes is expected to be a promising method for producing rice cultivars with desired agronomical traits. However, before GM rice is released into the market, its potential risks must be assessed to ensure safety.

Metabolomics, in conjunction with other allergenicity and toxicity tests, is critical in risk management. However, while a survey of the entire metabolome is required for detailed risk analysis, the number of detectable metabolites by currently available metabolomic techniques is limited.

Similarly, because scientific procedures or guidelines for assessing potential risks using metabolome data have not yet been established, ensuring substantial equivalence can be difficult, if not impossible. These drawbacks indicate that further improvement in metabolomics is essential for more comprehensive profiling and interpretation of rice metabolites.

3.7. METABOLOMICS FOR IMPROVEMENT OF FRUITS

Metabolomics is a rapidly growing discipline that has recently been used to assess the impact of fruit post-harvest processes and to identify the pathways involved. So far, most applications have been limited to metabolic fingerprinting to differentiate between phenotypes and metabolic profiling to identify biomarkers for post-harvest disorders and fruit spoilage.

Through a complete understanding of the metabolome, metabolomic approaches may also aid in the development of new strategies to improve the flavor quality and health-related attributes of fruit.

The knowledge base derived from metabolomic fruit studies has the potential to aid in the development of intelligent management decisions. Metabolomics has the potential to predict the impact of post-harvest operations on the physiology and quality of fruit.

However, scale-up studies and validation under real-world conditions are required for commercial applications. This section examines recent developments in the use of metabolomics to improve the quality of harvested fruit.

Metabolomics research has provided new insights into fruit biology, particularly ripening and quality. Tomatoes (*Solanum lycopersicum*) are high in carotenoids, antioxidants, and flavonoids. Metabolite segregation patterns of 50 tomato cultivars showed close agreement with fruit size segregation.

By plotting a correlation with the fruit transcriptome, the metabolome can be used to dissect the ripening event. Through a genome-wide metabolic survey, the metabolome can be used to elucidate diverse and differential biochemical pathways that exist in the fruits of tomato ILs and Ecotypes, as well as the ancestral species.

The peel and flesh of the apple (*Malus spp.)* contain beneficial nutrients, including antioxidants, which reduce the risk of chronic diseases such as asthma, cancer, cardiovascular disease, and diabetes. Apple fruit metabolite content is used to distinguish commercially important cultivars.

The cultivar 'Golden Delicious,' for example, has a high concentration of myo-inositol, sugars, and succinic acid, whereas the cultivars 'Red Delicious' and 'Fuji' have a higher concentration of triterpene/sterols, flavonoids, phenolic acids, stearic acid, anthocyanins, and carbohydrates.

Figure 3.3. 'Golden Delicious' apples.

Source: Image by Wikimedia Commons

The 'Fuji' fruit peel extract contained high levels of carbohydrates, including glucose and sorbitol, and was significantly different from the 'Red Delicious' and 'Granny Smith' fruit peel extracts, which contain high levels of unsaturated fatty acids (oleic and linoleic acids).

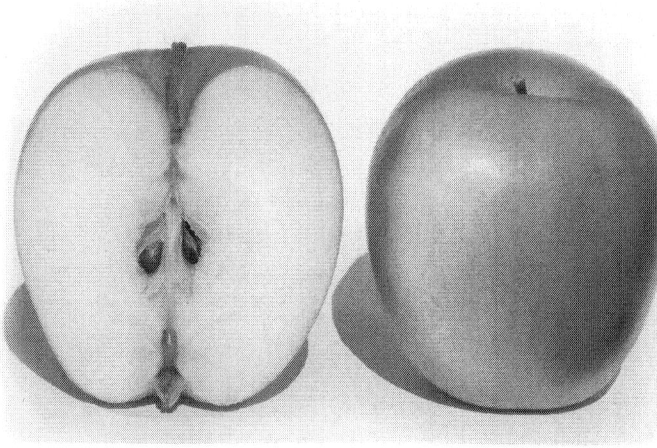

Figure 3.4. Granny Smith apples.

Source: Image by Wikimedia Commons

A recent study on metabolic profiling of the apple fruit revealed the spatial distribution of sugars and organic acids between fruit layers. The browning of apple fruits during storage renders them unmarketable, thus exerting an adverse impact on the apple industry.

The metabolomics study on stored apple fruits revealed a difference in the level of primary metabolites over time. Increased levels of mannose and xylose during post-harvest indicated hemicellulose breakdown in the cell wall, which promotes fruit senescence. The research found a link between metabolic regulation during post-harvest storage, cellular respiration, and stress.

The Kiwifruit (*Actinidia Lindl. spp.*) has gained popularity in international markets in recent years due to its distinct appearance and health-promoting nutrients such as vitamin C and fiber. During the development and ripening of Kiwifruit, 51 metabolites were detected.

The content of soluble sugars and ascorbate changes significantly during ripening, determining fruit quality and taste. Thus, the quality and flavor of Kiwifruit can be subjected to improvement by targeting the metabolites that can render consumer acceptance.

Figure 3.4. Kiwifruit.

Source: Image by Wikimedia Commons

In Kiwifruit, the application of synthetic cytokininN-(2-chloro-4-pyridyl)-N′-phenylurea significantly increases fruit size and affects the processes of ripening by altering the accumulation pattern of metabolites such as amino acids, sugars, organic acids, etc.,

The composition of metabolites such as organic acids, sugars, vitamins, flavonoids, and carotenoids determines the quality and flavor of orange (*Citrus spp.*) fruit.

The metabolomics analysis of the orange bud mutant 'Hong Anliu' (which accumulates larger quantities of lycopene and is sweeter than the natural type) revealed 130 metabolites, including acids, sugars, flavonoids, alkaloids, limonoids, coumarins, amino acids, and plant hormones.

At the ripe stage, higher quantities of soluble sugars and lower levels of organic acids, as well as differential levels of flavonoids, controlled the flavor and taste of 'Hong Anliu' sweet orange.

The infection of*CandidatusLiberibacter asiaticus*, the causal agent of Citrus Huanglongbing disease, decreases juice quality. The infection causes a significant drop in glucose, fructose, sucrose, and amino acids such as alanine, arginine, isoleucine, leucine, proline, threonine, and valine while increasing citrate and phenylalanine levels.

Heat treatment of fruit is extensively utilized to prevent fruit infection during post-harvest storage, according to metabolomics research.

In a study, heat treatment reduced the content of organic acids and amino acids in a study, but it increased the accumulation of metabolites like 2-keto-D-gluconic acid, tetradecanoic acid, oleic acid, ornithine, succinic acid, myo-inositol, glucose, fructose, sucrose, and turanose, which reduces the risk of post-harvest infection.

ABA has recently been discovered to regulate citrus cuticular wax biosynthesis during fruit development.

Fruit setting in grapes (*Vitis vinifera*) is influenced by a number of metabolites and is controlled by hormones and sugar metabolism pathways that are reprogrammed. The impact of grape metabolite content on geographical dispersion has been thoroughly studied.

Grapes cultivated in high-sun-light-low-rainfall areas had higher quantities of sugars and amino acids, as well as Na and Ca, and lower levels of organic acids, implying that external variables play a role in grape fruit quality.

The ripening processes are regulated by the metabolite quantity in grape berries, which is stage-specific and cultivar dependent. The main polyphenols found in grapes that impact the quality of drinking wine are stilbenes.

The MS analysis of grapes allowed the detection of several bioactive stilbenes like ampelopsin H, caraphenol, isohopeaphenol, trans-resveratrol, Z- andE-astringin, piceatanno, Z- andE-piceid, B pallidol and pallidol-3-O-glucoside and parthenocissin A.

The study was centered on the polyphenolic content of the grape and identified up to 450 compounds, including anthocyanin, glycoside aroma precursors, flavanols and procyanidins, flavones, and flavanones, phenolic acids, and stilbenes. Specifically, allowed the identification of several 100 compounds, which were used to build a new database of putative compounds (Grape Metabolomics).

Pear (*Pyrus communis*) is a member of the *Rosaceae* family and is known for its unique 'melting' texture. Pears are grown in Japan in great quantities. During the development and ripening of pear fruit, a metabolomics study revealed a difference in the accumulation of about 250 compounds.

Sugars (e.g., sucrose), sulfur-containing amino acids, phytohormones like ABA, and brassinosteroids accumulated as pear fruit ripened. Abscisic acid, auxin, brassinosteroids, gibberellins, jasmonic acid, and salicylic acid were among the 15 phytohormones found.

The metabolites (amino acids and organic acids) grow significantly during the flowering stage, then decrease during fruit development.

3.8. METABOLOMICS FOR IMPROVEMENT OF LEGUME CROPS

The legume crops such as chickpea (*Cicer arietinum*), common bean (*Phaseolus vulgaris*), cowpea (*Vigna unguiculata*), faba bean (*Vicia faba*), lentil (*Lens culinaris*), lupin (*Lupinus luteus*), mungbean (*Vigna radiata*), pea (*Pisum sativum*), peanut (*Arachis hypogaea*), pigeonpea (*Cajanus cajan*), and soybean (*Glycine max*) have a great contribution in the provision of nutrition, food security, and environmental sustainability.

Despite this, a majority of them are grown in a marginal environment, where they are subjected to abiotic (e.g., drought, heat, cold, salinity, waterlogging, heavy metal toxicities, etc.) and biotic stresses (e.g., anthracnose, bean rust, bacterial blight, Fusariumwilt, etc.), hence putting a limit on their productivity.

In addition, these environmental conditions have a severe effect on the rhizobia-legume symbiosis, which contributes to ~45% of nitrogen required for agriculture.

Therefore, for the development of varieties of superiority with enhanced stress tolerance, it is highly imperative to have an understanding of the stress response mechanisms found in legumes which includes changes in gene expression and the subsequent variations in the transcriptome, proteome, and metabolome.

The availability ofhigh-throughput, low-cost next-generation sequencing (NGS) tools, as well as high-throughput genotyping technologies, has aided the creation of enormous genomic data for model and crop legumes.

The contribution of forage and grain legumes is 27% of the total world's primary crop. Solely, the grain legumes cater to 33% of required human dietary protein, hence leading to a contribution to global food security and environmental sustainability.

Making the exception of a few extensively investigated model legumes, metabolomics studies in other legumes remain limited. The studies in model legumes present a decline in oxylipins as an effect of rhizobial node factor (Nod) inMedicagoand metabolic adjustments of shoot constituent in salt-tolerantLotus species for the survival of the same.

Salinity and anoxia are stress conditions that cause an accumulation of alanine, and its biosynthesis co-substrates glutamate and GABA and succinate in soybean. Genes involved in nitrogen-fixing and fermentation in the root were also found to have differential expressions.

Surprisingly, during waterlogging, there was a negative association between amino acids derived from glycolysis and the TCA cycle; multiple TCA cycle enzymes were stimulated.

Similarly, a study of metabolic changes associated with flooding stress in soybeans identified 81 mitochondria-associated metabolites, implying an increase in concentrations of metabolites involved in respiration and glycolysis such as amino acids, NAD, and NADH, as well as a decrease in free adenosine triphosphate (ATP).

The metabolite phenotyping of four different Mediterranean lentil accessions revealed a decrease in the TCA cycle and glycolytic pathway intermediates under drought and salinity conditions. Essentially, the study yielded metabolite markers for particular stress, such as threonate, asparagine/ornithine, and alanine/homoserine for NaCl, drought, and salinity, respectively.

Another study that looked at the effects of water scarcity on *Lupinus albus* found that the plant stem served as a sugar and amino acid storage organ. Importantly, the stem stelar region of tolerant plants accumulated substantial levels of metabolites such as asparagine, proline, sucrose, and glucose.

This points to the need for plants to reorganize their nitrogen and carbon metabolic pathways in order to withstand salt stress. Pinitol (sugar alcohol, osmoprotectant) levels in the tolerant soybean plant increased consistently in both normal and drought-stressed circumstances. Similarly, in reaction to water stress, tolerant soybeans accumulated sugar, free amino acids, and soluble proteins.

3.8.1. Metabolomics Databases for Legumes

With transcriptomics and proteomics data, metabolomics approaches also generate huge datasets that require specialized data mining and bioinformatics tools; integrating functional genomics data into a systemic approach, such as mathematical modeling of biological systems, is critical for completely studying biological components.

Because of developments in bioinformatics tools, the processing, mining, and analysis of metabolomics data have vastly improved. Several databases have been created to analyze plant metabolomics data in this circumstance.

A metabolic route reconstruction, for example, was used to create the Medicago pathway database MedicCyc, which includes over 250 pathways with linked genes, enzymes, and metabolites. The database includes isoflavonoid, lignin, and triterpene saponin biosynthetic routes for Medicago that were added or updated based on literature and in-house experience.

The design of the MedicCyc is created in such a way as to visualize functional genomics datasets from Medicago within the biological context of metabolic pathways. It has also been believed that this is best achieved through the visualization of data within the biological context of metabolic pathways in legumes.

These pathways were engineered in order to enable the correlated visualization of integrated functional genomics data. Soybean Knowledge Base (SoyKB), which is another database, has been reported to be a comprehensive resource for the soybean translational genomics, which contains integrative information on soybean genomics, transcriptomics, proteomics, and metabolomics.

It is a web resource that would not only be useful for soybean translational genomics but also for legume crop improvement programs. In addition, a database for plant metabolomics, PlantMetabolomics.org (PM), was developed. The metabolomics data generated from *Arabidopsis* is represented by this database through the integration of experiments compiled from different platforms with visualization tools.

Plant Metabolomics has been widely utilized for examining, visualizing, and downloading plant metabolomics data and well-annotated metabolomics datasets, which has been beneficial in establishing metabolomics as a functional genomics technique in legumes.

Legume crops are commonly grown in the semi-arid tropics, where their production is threatened by a variety of abiotic and biotic stressors. Under natural field conditions, the crops are susceptible to not just one but a variety of stresses at any particular time.

Crop development efforts using resourceful methodologies such as integrated "omics" approaches to understand stress responses at the molecular level such as cellular mechanisms, signaling pathways, and biochemical processes are required to sustain legume cultivation in these settings.

Model plants like *Arabidopsis* have been the focus of many of these investigations. For example, transcriptomics and the metabolomics data revealed a contradicting pattern of defense response in Arabidopsis subjected to a combination of drought and heat stress.

It could be found that proline (Pro) was accumulated in response to drought stress, but during the combination of stresses, Pro was replaced by sucrose as osmoprotectants. Studies like these need to be undertaken in legume crops such as chickpea, pigeonpea, and peanut, which are of agronomic value, particularly in the semi-arid tropics.

3.9. METABOLOMICS FOR IMPROVEMENT OF GM CROPS

Metabolomics, or the unbiased and nontargeted examination of the metabolome in organic tissue, has been proposed as a useful approach for investigating the metabolite profiles of genetically modified (GM) crops.

Before GM crops can be put on the market, they must undergo a thorough evaluation. The potential benefits and drawbacks of using nontargeted metabolomics-based analytical techniques in this procedure are discussed.

GM crops are currently widely used all over the world. According to the International Service for the Acquisition of Agri-Biotech Applications (ISAAA), biotech crops, including GM crops, were grown on 160 million hectares of arable land in 2011.

Metabolism is referred to as the processes involved in maintaining life, such as the synthesis and breakdown of proteins, nucleic acids, and carbohydrates. It provides a glance at the present biochemical status, which includes important nutritional and toxicological characteristics. The metabolite composition is further reported to have a close association with the organism's phenotype.

The reason why metabolomics comes in handy is when there is a need to investigate the metabolic composition of GM crops. The application of metabolomic technology could generate a database of metabolites in both GM crops and traditional varieties.

For instance, the metabolomics approach was employed to assess the chemical composition of GM tomatoes in order to compare the modified crops with the traditional varieties. The researchers utilized GM tomatoes that overexpress a foreign gene that codes for miraculin, a glycoprotein common in tropical plants but not in tomatoes.

In the tomato cultivars studied, the MS-based multiple platforms discovered 86 percent of the overall chemical diversity. The "similarities" and "differences" between GM tomatoes and six traditional cultivars, including the control line Moneymaker, were then evaluated using a statistical approach for "proof-of-safety" rather than "proof-of-hazard."

The results signified that the GM tomatoes had a reproducible metabolic signature; moreover, more than 92% of the compounds showed an acceptable variation in both green and red stages of the tomato, highlighting a striking similarity of the GM tomatoes with that of the control line Moneymaker in terms of their metabolite profiles.

Furthermore, the metabolite profiles obtained from two independent experiments were compared. The quantities of the most often changed metabolites in GM tomatoes, such as proline, 4-hydroxy-proline, spermidine, asparagine, arginine, serine, and inositol-1-phosphate, were assessed across all growth conditions in the study.

Genetic alteration had no effect on the expression of these metabolites, and they were not linked to the expression of foreign genes. This method could be beneficial for comparing GM crops to traditional crops in terms of metabolomic equivalency.

A vast number of novel transgenic crop types with desired features are being launched into the worldwide market at a rapid pace. Metabolomics is being used to analyze the safety of GMOs, and it provides useful information about the associated metabolite changes as a result of gene modification.

As this analysis shows, comprehensive, unbiased metabolic profiling or fingerprinting of plants has accelerated dramatically in recent years.

The chemical complexities of the plant metabolome, as well as the wide dynamic range of concentrations, are significant metabolomics issues. Advances in analytical platforms have played a major role in unraveling possible GM effects at the molecular level to address these problems.

In order to examine natural metabolome variability, metabolomics research (e.g., comparing GM crops to their wild-type parent lines) is frequently paired with diverse culture circumstances, from numerous geographical areas, multiple years, different growing seasons, etc.

Environmental variables create bigger significant alterations in metabolome composition than genetic modifications, according to the findings.

Although the applications of metabolomics are yet to be validated within the regulatory framework for food safety assessment, it can still be considered a significant approach and a key strategy to screen compositional changes, which increases the possibility of detecting unintended effects associated with genetic modification in GM plants.

However, as proposed by the new Foodomics strategy, the large amount of data generated in metabolomics studies must be processed, integrated, and interpreted together with data generated by other high-throughput technologies such as proteomics and transcriptomics in order to understand the biological significance and impact of the detected changes. Despite the use of a variety of omics technologies to investigate molecular changes in GMOs, none of these researchers have found significant connections due to a lack of appropriate integrative techniques.

In this regard, the development of relevant Systems Biology methodologies that provide new ways to integrate and synthesize omics data will aid the applicability of global approaches that can reveal new links that would otherwise be impossible to discover.

Nevertheless, before data integration processes can be used to analyze GMOs in a practical way, substantial work must be spent on developing effective integrative statistical approaches and adequate computational systems for characterizing molecular systems and integrating omics databases.

In this context, it will be critical to develop and implement projects to provide tools and databases that allow for the exchange of metabolite data obtained from metabolomics research.

3.10. METABOLOMICS FOR IMPROVEMENT OF CEREAL CROPS

Corn, often known as maize, is an American native and one of the world's most significant cereal crops. Corn is a staple food in many Mesoamerican, South American, and African countries, as well as a significant ethnic and cultural symbol for indigenous peoples in these areas.

Furthermore, this grain is used to make feed, fuel, and as a raw material for a variety of industrial products. In 2018, global corn output was estimated to be over 1147 million tons, with the United States, China, and Brazil as the top three producers.

Corn was domesticated from the wild annual teosinte in a single event around 9000 years ago in Mexico. Corn underwent major morphological transformations in the plant and inflorescence architecture during the domestication process, allowing it to adapt from its tropical origins to temperate and other different environmental circumstances.

Corn, both rudimentary and early refined, expanded across North and South America, adapting to new habitats and applications, giving rise to the concept of "landraces or races." Landraces are open-pollinated populations that are highly heterozygous and diverse.

"A collection of related individuals with enough traits in common to facilitate their designation as a group. Anderson and Cutler first proposed the definition of "race" as "a group of related individuals with enough characteristics in common to permit their recognition as a group."

Salazar et al. defined landraces or local races as the germplasm that has been cultivated for hundreds of years under traditional agriculture systems and which represent a high genetic diversity. Native farmers chose this material for its better adaptability to certain habitats, yield, nutritional value, cultural value, and tolerance to biotic and abiotic pressures.

Landraces are genetically heterogeneous populations that do not have access to professional crop improvement. Corn was brought to other continents, including Europe, Africa, and Asia, with varying patterns of adaptability and breeding, following the arrival of Europeans in America.

With the advancement of more sophisticated breeding techniques, high-yielding inbred corn lines (formed by self-pollination) may now be produced from landraces, and subsequently, hybrids or commercial varieties can be generated from crossed inbred lines.

Teosinte, landraces, and inbred lines are all terminology used to describe the existing corn germplasm, which are essential genetically diverse resources for crop improvement and food security.

Figure 3.5. Corn seed germplasm.

Source: Image by Flickr

Metabolomics is a valuable tool for identifying significant bioactive phytochemicals in corn biodiversities, such as phenolic and carotenoid components. For a better understanding of the underlying molecular pathways involved in the manufacture of functional metabolites, metabolomic data must be combined with data from other high-throughput omics technologies.

The use of novel phenotype-genotype molecular techniques for maize breeding, such as mGWAS, could hasten the development of unique functional corn-derived foods with higher levels of health-relevant chemicals like anthocyanins and carotenoids.

Corn landrace genetic diversity is an important source of candidate genes for modern breeding projects aimed at developing healthier foods in more resilient and sustainable ecosystems.

Comprehensive molecular characterization of global maize landrace diversity utilizing an integrated multi-omics method, including metabolomics, is a major problem. To identify and use suitable corn germplasm for various breeding applications, high-quality phenotypic data is required.

3.11. CONCLUSION

This chapter discussed the different platforms to gather metabolomic data. In this chapter, the concept of gene identification is also discussed. It discussed the various approaches to collate metabolite, phenotypic, and genotypic data. Towards the end of the chapter, it discussed the metabolomics for the improvement of fruits, legume crops, GM crops, and cereal crops.

REFERENCES

1. Clish, C., 2015. Metabolomics: an emerging but powerful tool for precision medicine. *Molecular Case Studies*, [online] 1(1), p.a000588. Available at: <https://www.ncbi.nlm.nih.gov/pmc/articles/PMC4850886/> [Accessed 8 June 2022].
2. Dr. Priyom Bose, P., n.d. *Gene Identification Tools in Bioinformatics*. [online] News-Medical.net. Available at: <https://www.azolifesciences.com/article/Gene-Identification-Tools-in-Bioinformatics.aspx> [Accessed 8 June 2022].
3. Frank, T. and Engel, K., 2013. Metabolomics of genetically modified crops. *Hot Topics in Metabolomics: Food and Nutrition*, [online] pp.18-28. Available at: <https://www.futuremedicine.com/doi/10.4155/ebo.13.323> [Accessed 8 June 2022].
4. GálvezRanilla, L., 2020. The Application of Metabolomics for the Study of Cereal Corn (Zea mays L.). *Metabolites*, [online] 10(8), p.300. Available at: <https://www.ncbi.nlm.nih.gov/pmc/articles/PMC7463750/> [Accessed 8 June 2022].
5. Kumar, R., Bohra, A., Pandey, A., Pandey, M. and Kumar, A., 2017. Metabolomics for Plant Improvement: Status and Prospects. *Frontiers in Plant Science*, [online] 8. Available at: <https://www.frontiersin.org/articles/10.3389/fpls.2017.01302/full> [Accessed 8 June 2022].
6. Oikawa, A., Matsuda, F., Kusano, M., Okazaki, Y. and Saito, K., 2008. Rice Metabolomics. *Rice*, [online] 1(1), pp.63-71. Available at: <https://thericejournal.springeropen.com/track/pdf/10.1007/s12284-008-9009-4.pdf> [Accessed 8 June 2022].
7. Oms-Oliu, G., 2012. Using metabolomics to improve the quality of harvested fruit. *CAB Reviews: Perspectives in Agriculture, Veterinary Science, Nutrition and Natural Resources*, [online] 7(047). Available at: <https://www.cabi.org/cabireviews/review/20123289339> [Accessed 8 June 2022].
8. Simó, C., Ibáez, C., Valdés, A., Cifuentes, A. and García-Cañas, V., 2014. Metabolomics of Genetically Modified Crops. *International Journal of Molecular Sciences*, [online] 15(10), pp.18941-18966. Available at: <https://www.mdpi.com/1422-0067/15/10/18941/htm> [Accessed 8 June 2022].

9. Tyagi, S., Singh, U., Kalra, T. and Munjal, K., 2010. *Applications Of Metabolomics - A Systematic Study Of The Unique Chemical Fingerprints: An Overview*. 3rd ed. [ebook] International Journal of Pharmaceutical Sciences Review and Research. Available at: <https://www.globalresearchonline.net/journalcontents/volume3issue1/Article%20019.pdf> [Accessed 8 June 2022].

CHAPTER 4

PLANT PHOSPHATE NUTRITION AND ENVIRONMENTAL CHALLENGES

CONTENTS

Food production has been practiced at farms where animals and plants were raised together since prehistoric times. However, as part of the Green Revolution, synthetic fertilizers became an integral aspect of modern agriculture, resulting in a progressive fragmentation of cropping and animal husbandry. As a result, fertilizer usage has climbed dramatically in the last half-century. Indeed, changes in nutrient cycles have created an environmental problem that has resulted in lasting damage to natural ecosystems, despite the fact that the genuine needs for food security for a growing human population are reasonable. The provision of phosphorus (P) for plant nutrition is one of the most pressing issues in agriculture, and it necessitates a closer examination from a variety of perspectives.

4.1. INTRODUCTION

Phosphorus is an essential nutrient in plants since it is a component of several major plant structural compounds and serves as a catalyst in the conversion of many key biochemical reactions. Phosphorus is known for its ability to capture and convert solar energy into beneficial plant chemicals.

Phosphorus is a necessary component of DNA, which serves as the genetic "memory unit" for all living things. It's also a part of RNA, which reads the DNA genetic code and uses it to make proteins and other substances that are necessary for plant structure, seed yield, and genetic transfer. Both DNA and RNA have phosphorus links that connect their structures.

Phosphorus is an essential component of ATP, plants' "energy unit." ATP is a phosphorus-containing molecule that is formed during photosynthesis and processes from seedling growth to grain production and maturity. Phosphorus (P) is an element that occurs naturally in the earth's crust, water, and all living beings.

Phosphorus (P) is one of sixteen elements required for plant growth. Because Virginia's soils are naturally poor in phosphorus, most cropping systems on these soils require phosphorus supplementation to enhance production potential. On soils that are naturally poor in phosphorus and soils that have been depleted due to crop removal, research has shown that applying phosphorus fertilizer boosts crop growth and yields.

The most common application of phosphorus in agriculture today is crop fertilization. Despite the well-documented economic benefits of phosphorus fertilization on crop productivity, too much of a good thing can be harmful. Excessive phosphorus in the soil poses a risk to water quality.

The goals of this bulletin are to improve understanding of the impact of phosphorus in agricultural systems on water quality and to present strategies for phosphorus management that are both cost-effective and environmentally friendly.

Phosphorus is an important macronutrient for crop growth and development. It takes a role in the production of phospholipids and nucleotides, both of which are structural components of cells and membranes.

Membrane permeability, photosynthesis, respiration, glycolysis, redox processes, and signal transduction are all affected by it (mechanisms that enable plants to obtain information from the environment and adapt themselves to its constant changes).

In summary, receiving proteins, such as the plant–fungus contact signal, are required for a response between cells and growth regulators, which frequently occurs under P-deficiency situations. P is also involved in lipid metabolism, carbohydrate transportation, and potential osmotic maintenance.

A decrease in bioavailable P, on the other hand, causes alterations in its stoichiometric interactions with other elements such as carbon (C) and, particularly, nitrogen (N) (N). P is essential in agriculture as a basic fertilizer since it is a dynamic element in nature.

Soil parameters such as pH, salinity, high concentrations of hazardous elements, interaction with micronutrients, runoff, leaching, total P concentration, soluble (and thus assimilable for plants) P pool, organic matter content, redox potential, soil structure and texture (particularly clay content), mineralogical composition, enzymatic activity, P-solubilizing microorganisms, and mycorrhizas all influence P dynamics.

Low available/labile P concentrations in the soil necessitate high P fertilizer applications, despite the fact that this practice can have major environmental consequences.

Recent investigations have found what is known as diffuse agricultural pollution of P, which is water pollution caused by a wide range of human activities with no evident point of entry into receiving watercourses. This pollution is primarily caused by crop fertilizers that are applied to the soil.

Crops take P from the soil through diffusion, but the element's diffusion coefficients are modest, and its concentration in the soil solution is limited. As a result of the low mobility of phosphate in the soil, it must be administered in close proximity to plant roots, either in bands (particularly for six-month crops) or in a half or full circle.

In contrast to its action in the soil, P is extremely mobile within the plant. In old leaves, P shortage gives a distinctive coloring that ranges from orange to reddish tones. This is related to a decrease in chlorophyll biosynthesis and an increase in pigment production, such as anthocyanins.

Soluble P is delivered to all growth points through the xylem, depending on P content, which is usually between 0.1 and 0.3 g P kg1 for most crops (grasses and agro-industrial crops, including fruit, legumes, and others).

Although P is essential for crops, its dynamics are intertwined with those of other elements. Because P is a macronutrient that inhibits primary production in many ecosystems, removing C from the atmosphere could necessitate a significant amount of P. Spohn demonstrated that organic C storage in mineral soils sequesters a significant amount of P.

Mineralization of organic C and the production of microbial necromass rich in P, as well as significant sorption of organic P to mineral surfaces that impede its mineralization, are the causes of the high enrichment of organic P in soils.

The lack of phosphorus supply and availability continues to be a major constraint in biological nitrogen fixation and symbiotic relationships. This demand may be higher than that for the growth of the host plant's roots or shoots.

Figure 4.1. Phosphorus deficiency symptoms on leaves of winter.

Source: Image by Wikimedia Commons

Plants and rhizobia have distinct P requirements, with slow-growing rhizobia being more tolerant of low P levels than fast-growing rhizobia. The nodules themselves are significant P sinks, and the availability of P has a big impact on nodulation and N2 fixation.

Water-soluble and neutral ammonium citrate soluble or citric acid-soluble phosphates are the most often used phosphatized fertilizers in agriculture. Apart from being used in traditional ways, simple superphosphates (SSP), triple superphosphates (TSP), monoammonium phosphate (MAP), and diammonium phosphate (DAP) are the most soluble P sources and provide the highest efficiency for crops, but they also contribute the most to environmental pollution.

Let's begin by discussing the role of P in living creatures and the evolution of adaptive mechanisms, focusing on molecular and genomic aspects. Following that, the cycle of transferring P between the physical and biological worlds will be outlined in order to demonstrate the extent to which contemporary farming methods have harmed the environment.

After then, potential solutions to the challenges that have arisen in industrialized agriculture will be examined.

To address integrative measures for sustainable agriculture, the need for introducing less-energy-demanding production and consumption methods for P provision, as well as the use of new generation fertilizers, particularly organic and biological fertilizers in combination with chemical P fertilizers is necessity.

These fertilizers are expensive to produce because they necessitate a lot of energy, as well as other soluble ingredients like sulphuric and nitric acid (imported raw material, which increases the selling price of the product even more). As a result, more study is needed on phosphatized controlled-release fertilizers, which can reduce fertilizer losses in the environment and improve fertilizer agronomic efficacy.

Materials like biochar or others with water-soluble P coatings with an insoluble layer, which creates a physical barrier to slow down the rate of P release, are being studied in this context. Graphene oxide, a nanomaterial being explored as a middle-term P transporter for crops, is one of the polymers used.

As a result, phosphorus is necessary for all plants' overall health and vitality. The following are some of the specific growth factors that have been linked to phosphorus:

- Stimulated root development
- Increased stalk and stem strength
- Improved flower formation and seed production
- More uniform and earlier crop maturity
- Increased nitrogen N-fixing capacity of legumes
- Improvements in crop quality
- Increased resistance to plant diseases
- Supports development throughout the entire life cycle

Figure 4.2. A picture of Biochar Compost.

Source: Image by Flickr

4.2. PHOSPHORUS DYNAMICS IN THE PLANT

Phosphorus is a necessary and irreplaceable component of life. It is one of the 16 important elements for plants' fundamental operations, and without it, no high-quality or plentiful crops would be possible. Nonetheless, the biomass potential of soil is constrained by the world's low P reserves.

P is an essential macronutrient for the synthesis of several cellular components, including nucleic acids, proteins, phospholipids, and ATP.

It is required for enzymatic regulation and the interpretation of metabolic signals, and because of its high mobility within the plant, signs of nutritional deficiency appear first in the older leaves, despite recent research showing that plants can be severely affected by a P deficiency for weeks without showing any visible symptoms on the leaves.

All photosynthetic processes impacted by P deprivation, on the other hand, appear to be totally reversible and can be restored in less than 60 minutes by administering orthophosphate to the leaf tissue.

Plants have evolved many ways to cope with P deprivation, which can be divided into two categories:

(i) Those that improve P absorption from the soil, and

(ii) Those that limit P reallocation in the vegetal organs.

Plants can alter their root system design to boost P uptake from soil by increasing surface area and developing several types of special roots called proteoid roots, radicles, and basal roots.

This leads to the exploration of bigger soil volumes, scavenging for P, and rhizosphere environment alteration to promote P mobilization. Roots may boost exudate output, acidifying the rhizospheric region and causing more H2PO4 to be produced, which is the preferred ion for acquired roots.

Exudation also includes the release of organic acids that compete with P for sorption sites, as well as enzymes like phosphatases and phytases that can hydrolyze various organic P forms. Plant root linkages with mycorrhiza, particularly arbuscular mycorrhiza, may improve plants' ability to acquire P via the hypha network, which improves spatial exploration.

Plants, on the other hand, reduce their use of P as their development rates slow down by storing sugars and anthocyanins, altering their glycolytic and respiratory pathways, and switching vegetal hormones.

Plants have also developed the ability to grow several types of special roots termed proteoid roots or radicles, as well as basal roots, to help them absorb P. The second is the formation of connections with mycorrhizas, particularly arbuscular mycorrhizas, which form a symbiosis with plant roots, increasing the spatial availability of P through the hyphae of these fungi.

The third technique involves altering the rhizosphere's environment to increase P mobilization. This includes the release of exudates by plant roots, acidifying the rhizospheric area so that there is more H2PO4 -, which is the easier form for plants to absorb (physiological factor) that is only found in acid soils, and acidifying the rhizospheric area so that there is more H2PO4 -, which is the easier form for plants to absorb (physiological factor) that is only found in acid soils.

The rhizosphere's exudation stimulates the synthesis of phosphatases by mineralizing the various forms of organic P in the soil.

Plants create a large number of secondary metabolites that are essential for survival in harsh environments. These include flavonoids and, in particular, anthocyanins, whose biosynthesis is influenced by a variety of intracellular signals as well as external stimuli. Anthocyanins build in plants as a result of P deficiency issues in crops.

Protein synthesis, which results from the variable expression of multiple stress-sensitive genes, is responsible for these adaptive responses. In order to understand the molecular mechanisms of adaptation, it is necessary to identify the genes and/or proteins implicated in adaptive responses to P deprivation.

Proteomic research is linked to strategies like the accumulation of anthocyanins in the shoot. The molecular processes for anthocyanin accumulation in induced plants under P suppression were determined using an iTRAQ-based proteome study using Arabidopsis.

The investigators determined that this flavonoid enhanced its metabolite content in the presence of P shortage, as evidenced by high levels of mRNA of seven proteins mapped in the experiment. This shows that a lack of P encourages the manufacture of anthocyanins, hence promoting their accumulation.

As a result, when plants experience early P deficiency, mature leaves take on violet, orange, or red tones due to an increase in anthocyanin concentration around the leaf region, resulting in a significant reduction in photosynthesis.

4.3. POTENTIAL ENVIRONMENTAL IMPACTS OF PHOSPHORUS

Phosphorus loss from agricultural soils can cause natural waters to become more fertile (Eutrophication), allowing algae and other aquatic plants to flourish faster. Phosphorus is usually the nutrient that prevents freshwater eutrophication.

The USEPA has proposed a total phosphorus limit of 0.05 ppm in streams that enter lakes and 0.1 ppm in flowing streams for controlling eutrophication (USEPA, 1986). Acceptable phosphorus levels in surface runoff from agricultural fields have yet to be determined.

Figure 4.3. Eutrophication: Algal blooms like this one are common in waters.

Source: Image by Flickr

Eutrophication has been linked to a variety of water quality issues. Algal blooms can kill fish and injure wildlife and cattle by lowering the oxygen concentration of the water (anoxia) or producing poisons. While high-value food fish, submerged macrophytes, and benthic creatures disappear, lakes may become dominated by algae and coarse, rapidly-growing fish.

Drinking water purification might become more expensive and complex as a result of eutrophication. Surface scum smells and a rise in insect pest populations are all signs of decaying algal biomass.

4.4. PHOSPHORUS DYNAMICS AND EFFECTS OF CLIMATE CHANGE

Phosphorus Dynamics and Climate Change Effects Temperature, water stress, wet/dry cycles, drought, and O2/CO2 concentration are all elements that affect P dynamics in the soil-plant system. These factors can have

Figure 4.4. A picture of Plant wastes & manures.

Source: Image by Pixabay

4.5.3. Inorganic Phosphorus

Apatite (the initial source of all phosphorus), complexes of iron and aluminum phosphates, and phosphorus adsorbed onto clay particles are all inorganic forms of soil phosphorus. These phosphorus compounds, as well as organic phosphorus, have very poor solubility, and only a small quantity of soil phosphorus is in solution at any given time. Most soils have less than a pound of soluble phosphorus per acre, with some soils having much less.

Soil solution phosphorus can be supplied quickly enough for optimal crop output with proper phosphorus fertilizer and good crop/soil management.

4.5.4. Soil Phosphorus Availability

Soluble phosphorus combines with clay, iron, and aluminum compounds in the soil or is rapidly transformed to less accessible forms via the process of phosphorus fixation, whether from fertilizer or natural weathering.

Because of these processes, phosphorus moves relatively little (less than an inch) in most soils, staying close to its source, and crops seldom absorb more than 20% of fertilizer phosphorus during the first cropping season following application. As a result, the leaching of soil phosphorus is minimal.

This fixed, residual phosphorus will persist in the rooting zone and become available to following crops over time. Soil phosphorus is lost mostly through soil erosion and crop removal.

4.6. FACTORS OF PHOSPHORUS AVAILABILITY

4.6.1. Soil pH

In calcareous soils with pH values around 8.0, phosphorus precipitates as weakly soluble calcium phosphates. Phosphorus precipitates as low-solubility Fe or Al phosphates in acidic environments. The pH range of 6.0 to 7.0 is where the most phosphorus is available.

Liming acid soils have a number of advantages, including these. The existence of H2PO4 ions, which are more readily taken by plants than HPO4+ ions, which occur at pH values above 7.0, is also favored by maintaining a soil pH in this range.

4.6.2. Balanced Crop Nutrition

The absorption of phosphorus from the soil is aided by an adequate supply of other plant nutrients. When ammonium forms of nitrogen are combined with phosphorus, phosphorus uptake is increased compared to when the phosphorus fertilizer is applied alone or when the nitrogen and phosphorus fertilizers are applied separately.

On neutral or basic soils, when soil phosphorus is available as calcium phosphates, sulphur applications often increase the availability of soil phosphorus.

4.6.3. Organic Matter

Organic matter-rich soils contain significant amounts of mineralized organic phosphorus (similar to organic nitrogen), which is available for plant growth. Organic matter not only provides phosphorus but also works as a chelating agent, combining with iron to prevent the development of insoluble iron phosphates.

Organic materials such as manure, plant residues, or green manure crops applied in large quantities to soils with high pH levels deliver not only phosphorus but also produce acidic chemicals, which increase the availability of mineral forms of phosphorus in the soil.

4.6.4. Type Of Clay

In soils, clay particles have a tendency to retain or fix phosphorus. As a result, fine-textured soils, such as clay loams, have a higher potential for phosphorus fixation than sandy, coarse-textured soils. Clays of the 1:1 type (kaolinite) have a higher capability for phosphorus fixation than clays of the 2:1 type (montmorillonite, illite, vermiculite).

Because kaolinitic clays are abundant in soils created under high rainfall and high temperatures, they have a substantially higher phosphorus-fixing potential than soils with 2:1-type clay. High temperatures and rains also increase the quantity of iron and aluminum oxides in the soil, which helps to fix the phosphorus that has been given to these soils.

4.6.5. Application Timing

The rate of phosphorus fixation in soil increases as the time spent in contact with soluble phosphorus and soil particles increases. As a result, fertilizer phosphorus utilization is often improved by applying the fertilizer immediately before planting the crop. This method works best in soils that have a strong phosphorus-fixing capacity.

Fertilizers can be sprayed many months before planting in coastal plain locations with little or no reduction in phosphorus availability to the crop. On soils with high phosphorus-fixing capacity, fertilizer banding is also significantly more likely to boost fertilizer phosphorus efficiency than on soils with poor phosphorus-fixing capacity.

4.6.6. Soil Temperature/Aeration/Moisture and Compaction

Low soil temperature and inadequate soil aeration reduce the plant's ability to absorb phosphorus. During cool weather, starter fertilizers containing water-soluble phosphorus are far more likely to boost crop development. Excessive soil moisture or compaction limits soil oxygen availability and the ability of plant roots to absorb phosphorus from the soil.

Compaction in the root zone limits aeration and pore space. Plant growth and phosphorus uptake are both slowed as a result of this. Compaction also reduces the amount of soil that plant roots may penetrate, limiting their total access to phosphorus in the soil.

4.6.7. Soil Test Phosphorus Levels

Crop responses to phosphorus fertilizer will be larger and occur more frequently on low-phosphorus soils than on high-phosphorus soils. Yields on soils with high p soil test levels, on the other hand, are usually higher. On high-testing soils, the sensitivity to phosphorus fertilizer is increasing, and it is critical to maintaining high soil phosphorus levels to promote optimum crop output.

4.7. ENVIRONMENTAL FACTORS AFFECTING PLANT GROWTH

The environment has a significant impact on plant growth and geographic distribution (where the plant may grow). A plant's development and/ or dispersion are limited by any environmental element that is less than optimum. In deserts, for example, only plants suited to low water availability may survive.

Most plant issues are induced by environmental stress, either directly or indirectly. In some circumstances, bad environmental conditions (such as a lack of water) directly harm a plant. Environmental stress can also weaken a plant, making it more vulnerable to disease or insect attacks. Light, temperature, water, humidity, and nutrition are all elements that influence plant growth.

It's essential to consider how these variables influence plant growth and development. One might be able to manage plants to fulfill one's demands if theyhave a rudimentary awareness of these aspects, whether it's for enhanced leaf, flower, or fruit output. One willbe better able to diagnose plant problems caused by environmental stress if one understands the roles of these components.

4.7.1. Phosphorus Deficiency in Plants

Phosphorus deficiency is harder to analyze than a lack of nitrogen or potassium. Crops ordinarily show no undeniable side effects of phosphorus lack other than a general hindering of the plant during early development. When a lack of visuals is remembered, rectifying in yearly crops might be past the point of no return.

A few crops, like corn, will quite often show strange staining when phosphorus is insufficient. The plants are normally dull, somewhat blue-green in variety, with leaves and stems becoming purplish. The level of purple is affected by the hereditary cosmetics of the plant, for certain mixtures show a lot more prominent staining than others.

The purplish variety is because of the gathering of sugars that leans toward the union of anthocyanin (a purplish-hued shade), which occurs in the leaves of the plant.

Phosphorus is profoundly portable in plants, and when lacking, it very well might be moved from old plant tissue to youthful, effectively developing regions. Subsequently, early vegetative reactions to phosphorus are frequently noticed.

Figure 4.5. Phosphorus deficiency in wheat.

Source: Image by Flickr

As a plant develops, phosphorus is moved into the fruiting region of the plant, where high-energy necessities are required for the structures of seeds and natural products. Phosphorus deficiency late in the developing season influence both seed advancement and ordinary harvest development.

The level of the aggregate sum of every supplement taken up is higher for phosphorus late in the developing season than for one or the other nitrogen or potassium.

4.7.2. Plant Nutrition

Plant nourishment frequently is mistaken for preparation. Plant sustenance alludes to a plant's requirement for and utilization of fundamental substance components. Treatment is the term utilized when these materials are added to the climate around a plant. A great deal should occur before a substantial component in manure can be utilized by a plant.

Plants need 17 components for typical development. Three of them- - carbon, hydrogen, and oxygen- - are tracked down in air and water. The rest are tracked down in the soil.

Six soil components are called macronutrients in light of the fact that they are utilized in moderately enormous sums by plants. They are nitrogen, potassium, magnesium, calcium, phosphorus, and sulfur.

Eight other soil components are utilized in a lot more modest sums and are called micronutrients or minor components. They are iron, zinc, molybdenum, manganese, boron, copper, cobalt, and chlorine.

The greater part of the supplements a plant needs are broken down in the water and afterward consumed by its foundations. As a matter of fact, 98% are ingested from the soil water arrangement, and something like 2% is really extricated from soil particles.

4.7.3. Nutrient Absorption

Whatever diminishes or stops sugar creation in leaves can bring down supplement assimilation. In this way, on the off chance that a plant is under pressure in view of low light or outrageous temperatures, supplement lack might create.

A plant's formative stage or pace of development likewise may influence how much supplements are ingested. Many plants have a rest (lethargic) period during part of the year. During this time, not many supplements are retained. Plants additionally may retain various supplements as flower buds create than they do during times of quick vegetative development.

4.7.4. Fertilizers

Fertilizers are materials containing plant supplements that are added to the climate around a plant. By and large, they are added to the water or soil, yet some can be sprayed on leaves. This technique is called foliar application.

It ought to be done cautiously with a low concentration, in light of the fact that a high manure fixation can harm leaf cells. The supplement, notwithstanding, necessities to go through the slender layer of wax (cutin) on the leaf surface.

By the way, fertilizers are not plant food! Plants produce their own food from water, carbon dioxide, and sun-oriented energy through photosynthesis. This food (sugars and starches) is joined with plant supplements to create proteins, chemicals, nutrients, and different components vital for development.

4.7.5. Temperature

Temperature impacts most plant processes, including photosynthesis, transpiration, respiration, germination, and flowering. As temperature expands (to a certain degree), photosynthesis, transpiration and respiration rise.

When joined with day length, temperature likewise influences the change from vegetative (verdant) to reproductive (flowering) development. Contingent upon the circumstance and the particular plant, the impact of temperature can either accelerate or revert back this progress.

4.7.6. Germination

The temperature expected for germination changes by species. By and large, cool-season crops (e.g., spinach, radish, and lettuce) develop best at 55° to 65°F, while warm-season crops (e.g., tomato, petunia, and lobelia) flower best at 65° to 75°F.

4.7.7. Flowering

In some cases, horticulturists use temperature in blend with day length to control flowering. For instance, a Christmas desert crop flower because of short days and low temperatures. To urge a Christmas cactus to flower, place it in a room with over 12 hours of darkness every day and a temperature of 50° to 55°F until flower buds form.

On the off chance that temperatures are high and days are long, cool-season yields, for example, spinach, will flower. Nonetheless, in the event that temperatures are excessively cool, the organic crop won't set on warm-season yields like a tomato.

4.7.8. Photosynthesis And Respiration

Thermoperiod alludes to day-to-day temperature change. Plants develop best when the daytime temperature is around 10 to 15 degrees higher than the evening time temperature. Under these circumstances, plants photosynthesize (anabolism) and respire (catabolism) during ideal daytime temperatures and afterward shorten their respiration around evening time.

In any case, not all plants develop best under similar reach among evening and daytime temperatures. For instance, snapdragons develop best at evening temperatures of 55°F; poinsettias at 62°F.

Temperatures higher than required increase in respiration, now and again over the pace of photosynthesis. In this way, photosynthates are utilized quicker than they are produced. For development to happen, photosynthesis should be more prominent than respiration.

Daytime temperatures that are too low frequently produce unfortunate development by reducing photosynthesis. The outcome is reduced yield (i.e., organic produce or grain developement).

4.7.9. Breaking Dormancy

A few plants that fill in chilly districts need a specific number of long periods of low temperature (lethargy). Knowing the time of low temperature expected by a plant, if any, is fundamental in inspiring it to develop to its true capacity.

Peaches are a great example; most development require 700 to 1,000 hours somewhere in the range of 32° and 45°F prior to breaking their rest period and starting development. Lilies need a month and a half of temperatures at or somewhat underneath 33°F prior to flowering.

Daffodils can be compelled to bloom by putting away the bulbs at 35° to 40°F in October. The chilly temperature permits the bulbs to develop. When moved to a nursery in midwinter, they start to develop, and blossoms are prepared to harvest in three weeks to about a month.

4.8. PHOSPHATE ION IMPORTANCE

P, as phosphate ion (Pi), is the most crucial component for all living creatures assuming significant parts in the designs of fundamental biomolecules, for example, nucleic acids, phospholipids, and phosphate-sugars, in practically all metabolic responses, including photosynthesis and respiration, in energy conveying atoms like ATP, ADP or NADPH and in the transduction of signs inside the cells.

To guarantee useful metabolic responses, Pi homeostasis should be held between 5 to 20 mM in the cytoplasm. Plants assimilate P just in its dissolvable inorganic type of Pi, H2PO4-or HPO42-, which occurs in the soil between 0.1 to 1 μ. In this manner, it is one of the most required supplements for plant development and improvement and is considered a significant restricting element in crop yield.

4.9. SOIL PI AND PLANTS UPTAKE

Most soils contain a lot of P compounds, going from 200 to 3000 mg/kg, found with the middle value of 1200 mg/kg. P intensifies in soil and contains a wide type of natural and inorganic structures. Be that as it may, just a little extent (by and large under 1%) is quickly accessible to plants as free Pi.

Most inorganic mixtures are strongly related to calcium (Ca) in basic soils or with iron (Fe) and aluminum (Al) in acidic soils.

Natural Pi represents 30 to 80 percent of soil P, among them, monoester P occurs dominatingly as cation subsidiaries of inositol hexakisphosphates (fundamentally as phytate), though sugar phosphates and diester phosphates (for example, nucleic acids and phospholipids) comprise just a little extent (~5%).

Factors that add to the accumulation and turnover of various types of in/natural P in soil are complex and constrained by different contending processes that have been the subject of a few surveys.

High centralizations of Pi are by and large found in the surface layer of soil profiles or in nutrient-rich patches. Plants generally produce a bigger number of roots in the surface soil than the soil.

For example, an examination of characteristics related to the pace of Pi take-up in wheat showed that root length thickness in the surface soil was the main quality for Pi uptake. Hence, drying the surface soil can cause stopped Pi take-up or a 'healthful dry season.'

Low dissolvability of Pi in water (0.5 mg/lit), slow dissemination paces of Pi in soil (10-12 to 10-15 m/s), and restricted limit with respect to recharging of soil Pi-systems are central points that add to its lack in plants. It is likewise affected by natural cycles, for example, the hydrolysis of ester bonds in natural Pi compounds by phosphatase proteins.

The take-up of Pi from the soil by plants relies upon both the pace of dissemination of Pi towards roots and the development of the root foundation to get to unexploited soil. Establishes quickly exhaust Pi in the soil systems so that its fixation at the root surface is assessed around 0.05-0.2 mm.

Albeit this lays out a Pi-dispersion slope from the rhizosphere to mass soil, a low Pi dissemination rate really restricts its take-up. It is accepted that appropriate use of the manures gives Pi, yet in addition, advances root development into unexploited soil.

4.9.1. Pi Uptake and Reallocation in Plants

Under experimental conditions, both high and low partiality Pi take-up components have been perceived in plants. Pi level is inside the micromolar range (1-10 μm), which compares to Pi fixations in most developed soils, the high-fondness carriers handle Pi take-up.

The Km for high-affinity carriers fluctuates from 1.8 to 9.9 μM. It is an energy-interceded co-transport process driven by protons created by a plasma layer H+-ATPase. Extra proof for the association of protons in Pi take-up comes from the utilization of inhibitors that disturbs proton angle across membranes, causing the concealment of Pi take-up.

Both experimental information and genome sequence assays demonstrate that plants have groups of Pi carrier genes. Present information recommends that members from the PHT1 Pi carrier family intervene move of Pi into cells, though members from the PHT2, PHT3, PHT4, and pPT families are engaged with Pi to move across inside cell and organelle layers.

Individuals from the PHT1 Pi carrier gene family have been recognized in a wide variety of plant species, including *Arabidopsis*, rice, *Medicago*, tomato, and soybean. Investigation of entire genome sequences of *Arabidopsis* uncovered a group of nine PHT1 carriers. Eight of them are expressed in roots, from which four are expressed in the epidermal cells.

Conversely, there is less overt repetitiveness in the elevated tissues. In rice, somewhere around 10 of the PHT1 carriers are expressed in roots.

Covering articulation designs have additionally accounted for the PHT1 Pi carriers in other plant species. The capacity of some PHT1 carriers has been broken down either by articulation in yeast Pi transport freaks or in plant cells. In *Arabidopsis* two Pi carriers, PHt1;1 and PHt1;4, intervene 75% of the Pi take-up limit of the root's structure in a large number of ecological conditions.

After take-up into the roots, Pi moves directly from the root surface to the xylem at around 2 mm/h and to different organs subsequently. Going into the xylem for significant distance movement to the shoot is worked with by one more systems of carrier-like proteins.

The majority of the consumed Pi by the roots is transported through the xylem to developing leaves. In Pi-deficient plants. stored Pi in more seasoned leaves is re-translocated to both more young leaves.

Also, developing roots, from where Pi can again be reused to the shoot. Subsequently, the take-up and assignment of Pi in plants require various transport systems that should work in show to keep up with a homeostatic degree of it all through the plant tissues.

4.10. EFFECT OF P FERTILIZERS ON CROP YIELDS

Soil scientists distinguish three types of Pi:

(1) Deposited Pi in the form of chemical compounds in the soil;

(2) Accessible Pi or free Pi ion in the soil; and

(3) Absorbable Pi or free Pi ion in the root's surroundings.

Despite the fact that the average application of P2O5 (19 kg/ha) was lower than the suggested amounts in the literature (30 kg/ha), crop yields increased significantly over the world. In Middle Eastern countries, for example, barley, wheat, and chickpea yields increased by 9 to 60% compared to controls.

When up to 90 kg/ha of P2O5 was applied to cereal crops, productivity indexes reached 11.1. Groundnut and mustard have productivity indexes of up to six, according to data on oil crops. Soluble Pi was added to soybeans in Indonesia, and the rise was up to tenfold. Such examples exist in the literature from the early nineteenth to early twentieth centuries, when the use of Pi fertilizers was becoming more prevalent.

The high yield is accompanied by the removal of Pi from the soil at a rate of 6 kg/ha for cereals and 20 kg/ha for other crops.

4.11. NEED FOR PI CONSERVATION AND RESTORATION

Since the mid-nineteenth century, alternative sources such as guano, deposited bird droppings on remote islands, have been used to replace local organic materials. By the end of the nineteenth century, guano resources had run out, and trade had shifted to PR. PR output has increased to around 140-160 Mt per year in recent years, allowing for the extraction of about 25 Mt elemental P each year.

Because PRs are non-renewable, the economically recoverable reserves are expected to be gone by the end of the century. The premise is that once high-quality and easily accessible Pi resources are limited, mining and processing will become outrageously expensive. The consumption of P has followed a Gaussian distribution curve since the 19th century, as seen in.

Assuming a total earth reserve of 3240 Mt of elemental P, production will peak at 29 Mt in 2033, as predicted by Cordell et al.

Although the exact timing is dependent on a number of factors, this is particularly concerning given the finite nature of non-renewable Pi reserves and the lack of a substitute for PRs.

Furthermore, one can expect that energy costs will rise in the future and that the average grade of Pi will fall, as it did between 1970 and 1996 when it fell from 15% to 13%. Because this dilemma is posing a silent danger to food security, globally coordinated action is required.

4.12. ENVIRONMENTAL CONCERNS ABOUT PI PRODUCTION

Vast volumes of soil are removed, and intensive washing with massive amounts of water occurs, resulting in significant changes in the environment of the mining sites and nearby areas. The most concerning problem, however, is the disposal of side products, which heavily contaminate the ecosystem near stacking sites.

Phospho - gypsum, a byproduct of the wet process, is deposited in gypsum stacks, which are huge geographical expanses. About 250 Mt of gypsum is generated annually to produce 50 Mt of P2O5, of which only 10% is used for other purposes (e.g., road building, cement, and housing), and the remainder is stockpiled.

Another difficulty for the Pi manufacturing sector is the presence of radioactive elements in side products. Uranium (20-300 ppm) and thorium (1 to 5 ppm) are the most abundant elements in PRs, as well as their decay products. This equates to 0.35 kg U2O8 per Mt P2O5, or 2100 tons of uranium wasted per year. In the wet process, about 80% of this amount and their decay products are concentrated within the gypsum waste materials.

These could range from 7 to 100 picocuries per gram of uranium, 11 to 35 picocuries per gram of radium, and 1.7 to 12 picocuries per gram of radon, while radioactivity varies depending on the original PR. The radioactivity of the fertilizers themselves could be as high as 33 pCi/g in the United States.

4.13. ENVIRONMENTAL IMPACTS OF PI FERTILIZERS CONSUMPTION

The world's ever-increasing population has put great pressure on farmers to produce larger agricultural fields and raise crop yields through improved mechanized technologies and fertilizer use. Such methods not only change the chemical and physical qualities of soils but also have a significant impact on macro and microflora.

Because of their polycrystalline structure and inclination to precipitate, Pi fertilizers are not always guaranteed to be taken up by plants. If there is no crop present, adding Pi fertilizer to the soil raises the soluble P for the first couple of years, but then it decreases to a constant level thereafter. Pi application may not be as effective as it once was, according to studies conducted in industrialized countries.

In addition to the impact of deposited Pi on soil structure and microbial flora, Pi leaching into water streams and lakes due to erosion of surface soil and organic matter causes massive eutrophication.

A portion of Pi inputs to surface and subsurface waters must be accounted for by discharged detergents, sodium triphosphates, and human or animal metabolic wastes. Algal blooms and anoxic events can occur as a result of eutrophication and the over-enrichment of aquatic habitats with nutrients, primarily Pi and nitrate, which can cause irreversible environmental harm.

Although some lakes have recovered once nitrogen supplies are lowered, lakes can stay eutrophic for years due to the recycling of enriched Pi from sediments.

In affluent countries like Japan and Germany, recycling Pi from sewage systems, precision farming, and agricultural and industrial treatment units has been used to limit such consequences. Despite its inferior efficacy, aluminum silicate has been proposed as a replacement for Pi in detergents.

To recover or utilize wasted Pi, communities have sought to crystallize Pi in wastewater as struvite (ammonium magnesium phosphate), separate urine, and use sterilized faeces.

With the assumption that PRs are the only high-value Pi reserves, Cordell et al. cautioned about the peak time of Pi production. Given the large volumes of insoluble Pi compounds that have accumulated in the soil, using PSMs as biofertilizers could be a simple solution.

Pi biofilters' simplified application method and comparable cost, on the one hand, and higher quantity and quality of yields, on the other, have already piqued farmers' interest.

4.14. THE EFFICIENCY OF P FERTILIZERS IN AGRONOMIC MANAGEMENT AND ENVIRONMENTAL IMPACT

The agronomic efficiency of such fertilizers is still insufficient. This is due in part to the low rate of nutrient absorption in most soils and in part to losses. To produce crops, almost 70% of the world's arable land needs a large amount of P. Plant absorption, on the other hand, is slow at short lengths of around 0.1 to 0.15 mm.

Other publications claim that Phosphorus mobility and concentration in soils are quite low when compared to other nutrients; phosphate diffusion coefficients in the soil range from 0.3 - 3.3 10^{-13} m2 s1, and its concentration in the soil solution is 0.02 ppm. Fertilizer given to a crop but not consumed by plants has severe environmental and economic implications.

Eutrophication-related pollution and groundwater or runoff water hypoxia are among the environmental costs. P enrichment as a result of human activities is one of the primary causes of eutrophication of surface waters, which results in an explosion of autotrophic organisms that has a significant impact on aquatic ecosystem quality.

As a result of this challenge, various technologies have been developed to provide phosphatized fertilizers with high agronomic efficiency in the soil, high yields, and little environmental effect. The capacity of a P source as a fertilizer to meet the needs of agricultural plants during their phenological phases (agronomic efficiency) and its solubility rate are both factors to consider when selecting a P source as a fertilizer.

When choosing P-based fertilizers, other factors such as the supply of other nutrients (synergy), pH, granulometry, mechanical resistance, segregation, density, hygroscopicity, saline index, and physical and chemical compatibility in the mixes can be taken into account. The agronomic efficiency of various phosphatized fertilizer sources, potential contamination, and new alternative fertilizer sources on the global market.

4.15. ENVIRONMENTAL CHALLENGES OF PHOSPHORUS

Phosphorus-related environmental challenges cover a wide range of topics. Some are caused by improper material use, while others are not. The most immediate environmental consequence of widespread phosphorus use in modern civilizations, eutrophication, has gotten a lot of attention. However, it is not the entire story, and other problems must be considered.

Here, the focus is on five phosphorus-related environmental issues: mineral reserve conservation, soil erosion and degradation of soil fertility, animal waste management, sewage, and detergent use, and eutrophication.

Phosphorus is a non-toxic nutrient for plants. Generally, the crop P requirements are much higher than actual crop P uptake. Plants only consume a small portion of the available phosphorus in the soil.

The amount of P that remains in the soil rather than being taken up by plant growth is referred to as residual soil P. That residual P contributes to the accumulation of soil phosphorus, specifically the accumulation of available phosphorus in the tillable soil layer.

The increase in available soil phosphorus caused by continuous P fertilizer additions results in excessive phosphorus levels over time. As a result, the soil becomes phosphorus-saturated. The soil P saturation percentage is calculated using the ratio (percentage) of available soil phosphorus to soil exchangeable aluminum, both of which are extracted and measured using the same Mehlich III extracting solution.

According to data collected from the Agricultural Laboratory's database in Fredericton, New Brunswick, from September 1999 to July 2001, 60% of agricultural soils in New Brunswick tested higher than 78 ppm phosphorus. Above 78 ppm P, the tillable soil is considered very rich in available phosphorus for most crops, particularly potatoes, forages, and cereals.

High P levels in the 0-15 cm surface soil layer correlate with high P saturation percentages in the soil. Under erosion conditions, the potential for dissolved and sediment-attached phosphorus to enter watercourses is high. When phosphorus enters a watercourse, it contributes to phosphorus enrichment. Water eutrophication is another term for this enrichment.

Water eutrophication promotes the growth and development of aquatic plants. More plants are growing, producing, and dying. The decomposition of those dead organic materials causes dissolved oxygen depletion, threatening the development of beneficial, aerobic aquatic life.

Nitrogen and phosphorus are naturally occurring nutrients in aquatic ecosystems. In addition, nitrogen is the most abundant element in the air we breathe. Nitrogen and phosphorus promote the growth of algae and aquatic plants, which provide food and habitat for fish, shellfish, and other aquatic organisms. However, when too much nitrogen and phosphorus enter the environment, which is usually caused by a variety of human activities, the air and water can become polluted. For several decades, nutrient pollution has impacted many streams, rivers, lakes, bays, and coastal waters, causing serious environmental and human health issues as well as economic impacts.

When there is too much nitrogen and phosphorus in the water, algae grows faster than ecosystems can handle. Significant increases in algae harm water quality, food resources, and habitats and reduce the oxygen required by fish and other aquatic life.

Algal blooms are large growths of algae that can severely reduce or eliminate oxygen in the water, causing illnesses in fish and the death of large numbers of fish. Some algal blooms are toxic to humans because they produce elevated toxins and bacterial growth, which can make people sick if they come into contact with polluted water, eat contaminated fish or shellfish, or drink contaminated water.

Figure 4.6. Algal bloom and destruction of a waterbody.

Source: Image by Wikimedia Commons

Nutrient pollution in ground water, which millions of Americans rely on for drinking water, can be harmful even at low levels. Nitrates, nitrogen-based compounds found in drinking water, are dangerous to infants.

Excess nitrogen in the atmosphere can lead to the formation of pollutants such as ammonia and ozone, which can impair our ability to breathe, reduce visibility, and alter plant growth. When excess nitrogen from the atmosphere returns to the earth, it can harm the health of forests, soils, and waterways.

Excessive nitrogen and phosphorus in the water can have a wide range of consequences for public health, the environment, and the economy. A high amount of nitrogen in the air can impair breathing, reduce visibility, and alter plant growth.

4.16. CONCLUSION

A lack of phosphorus in plants is more difficult to break down than an absence of nitrogen or potassium. Phosphorus is significantin plants, and while lacking, it may be moved from old plant tissue to actively proliferating sites.

A few crops like corn, will frequently show distinct staining when phosphorus is inadequate. The plants are typically dull rather blue-green in assortment, with leaves and stem becoming purplish. Temperature influences most plant processes, including photosynthesis, transpiration, respiration, germination, and flowering.

Composts are materials containing plant supplements that are added to the environment around a plant. Plants produce their own food from water, carbon dioxide, and sun-situated energy through photosynthesis.

This food is gotten together with plant nutrients to make proteins, synthetic substances, supplements, and various parts imperative for advancement. Thermo-period implies everyday temperature change. Plants grow best when the daytime temperature is around 10 to 15 degrees higher than the nighttime temperature.

REFERENCES

1. Ali, M., Samaeian, A., Sadegh, M. and Lohrasebi, T., 2012. Plant Phosphate Nutrition and Environmental Challenges. *Plant Science*, [online] Available at: <https://www.semanticscholar.org/paper/Plant-Phosphate-Nutrition-and-Environmental-Malboobi-Samaeian/8ca4af93dd03ec3c363a0dda84b3b4cc26114467> [Accessed 7 June 2022].
2. Ali, M., Samaeian, A., Sadegh, M. and Lohrasebi, T., 2012. Plant Phosphate Nutrition and Environmental Challenges. *Plant Science*, [online] Available at: <https://pdfs.semanticscholar.org/776c/bd94a3bfa22861864752197144196 4d14139 pdf?_ga=2.12308476.916554234.1651922490-376891579.1651922490> [Accessed 7 June 2022].
3. Griffith, D., n.d. *Phosphorus - Nutrient Management*. [online] Mosaic Crop Nutrition. Available at: <https://www.cropnutrition.com/nutrient-management/phosphorus> [Accessed 7 June 2022].
4. Liu, Y., Villalba, G., Ayres, R. and Schroder, H., 2008. Global Phosphorus Flows and Environmental Impacts from a Consumption Perspective. *Journal of Industrial Ecology*, [online] 12(2), pp.229-247. Available at: <http://www.indiaenvironmentportal.org.in/files/Global.pdf> [Accessed 7 June 2022].
5. Lizcano-Toledo, R., Reyes-Martín, M., Celi, L. and Fernández-Ondoño, E., 2021. Phosphorus Dynamics in the Soil–Plant–Environment Relationship in Cropping Systems: A Review. *Applied Sciences*, [online] 11(23), p.11133. Available at: <https://www.mdpi.com/2076-3417/11/23/11133> [Accessed 7 June 2022].
6. Mullins, G., 2009. *Phosphorus, Agriculture & The Environment*. [ebook] Virginia Tech. Available at: <https://efotg.sc.egov.usda.gov/references/public/va/PhosphorousAgEnv.pdf> [Accessed 7 June 2022].
7. OSU Extension Service. n.d. *Environmental factors affecting plant growth*. [online] Available at: <https://extension.oregonstate.edu/gardening/techniques/environmental-factors-affecting-plant-growth#:~:text=In%20some%20cases%2C%20poor%20environmental,%2C%20water%2C%20humidity%20and%20nutrition.> [Accessed 7 June 2022].

8. Sawyer, J., Creswell, J., and Tidman, M., n.d.*Phosphorus Basics | Integrated Crop Management*. [online] Crops.extension.iastate.edu. Available at: <https://crops.extension.iastate.edu/encyclopedia/phosphorus-basics> [Accessed 7 June 2022].

CHAPTER 5

IMPROVING NITROGEN AND PHOSPHORUS EFFICIENCY FOR OPTIMAL PLANT GROWTH AND YIELD

CONTENTS

MZs are areas in a field that have similar soil conditions and landscapes. Zones are homogenous if they have things in common, like similar crop yields, the same electrical conductivity (EC), and areas chosen by the producer. Some of the things that the attributes have in common are how they affect the environment, how well they use the inputs, and how much they can grow.

Figure 5.2. A researcher doing soil fertility analysis.

Source: Image by Wikimedia Commons

Researchers came up with a number of ways that could be used to define the edges of MZs. In geographic information systems (GIS), geospatial statistical analyses are used to group or combine geo-referenced data layers, such as soil color, electrical conductivity, yield, and topography, to define zone boundaries.

To figure out where the MZ is, people have used soil mapping units, remote sensing, topography, yield maps, soil, etc. Most of the delineation of the MZ is based on sources that are static and don't change (because of the effects of time on yields). Because they are static and don't change, they probably aren't the best way to account for all the different n requirements in a field.

5.3. USE OF TISSUE ANALYSIS FOR N MANAGEMENT

The crop's N status can be shown by the amount of N in critical states. Critical N is the least amount of nitrogen a plant needs to grow as much as possible at a certain time. When the corn plant first starts to grow and develop, there is a lot of N in the soil.

As the plant gets older, the amount of N in the soil goes down. The picture of this process is called "critical N dilution." The N nutrition index (NNIN) is the ratio of the actual N in the plant to the critical N set by experiments in the past.

If NNI is more than 1 or less than 1, it means that the crop can grow without limits or that it has grown enough. The NNI approach has been used with wheat (*Triticum aestivum L.*), grain sorghum (*Sorghum bicolor L.*), rapeseed (*Brassica napus L.*), rice (*Oryza sativa L.*), and grasses.

The appearance of critical N, which is thought to be caused by competition between corn plants in their early stages of growth, does not help us get a good idea of the crop's N status. An increase in crop biomass will cause the amount of N in the soil to go down. This is sometimes called "dilution."

5.4. SPATIAL VARIATION

Types of yields for commercial corn production are based on differences in things like soils, soil management techniques, production history, how water and nutrients move, and how space is different. Because of the differences in space, plants' needs for nitrogen (N), their susceptibility to stress, and how productive they are can change across a landscape.

Changes in the slope of the land in the middle of it, as well as the depth and drainage of the soil, can have a big effect on the amount of grain yield variation and the amount of corn grain yield.

Because of low water levels and the deposition of clay- and organic-rich soils into depressions and foot slopes in areas where corn is grown commercially, these landscape features have a higher level of N fertility than the rest of the landscape.

Since these nutrient-rich materials are low in OM, their movement downward has a noticeable effect on the soil in the upper landscape positions. This movement down also changes the concentrations of phosphorus (P) and potassium (K), which are easier to find in footholds and depressions. High rates of crop removal are a natural result of high rates of crop production in the past.

This could lead to P and K levels that are lower than expected. This suggests that, unlike OM, the redistribution and deposition of soil P and K may not be as strongly linked to changes in slope, suggesting resistance to movement. Crop stress could cause growth and yield to slow down. In some different kinds of landscapes, the seasons affect crops in different ways.

Changes in yield caused by different positions in the landscape during dry or wet growing seasons are made bigger. When compared to upland areas, landscapes with high levels of organic matter (OM) or that can hold a lot of water can be more resilient to the harsh conditions caused by droughts. If a lot of rain causes water to pool in depressions in the landscape, yields can go down.

5.5. FERTILIZER PLACEMENT AND TIMING

N application can make sure that crops with high NUE needs get the high level of N that they need. UAN (solutions of urea and ammonium nitrate) that are injected have better yields than UAN, which is spread on the ground, especially on landscapes with surface residue.

N loss can happen in a number of ways when broadcast UAN applications are used, such as when the ammonia in the urea part of the UAN evaporates and when N gets stuck in the surface residue of the landscape. Because of this, putting fertilizer below the soil's surface may be more effective.

In modern corn hybrids, the V7 growth stage is when about 15% of the total N is taken in, and 5% of the total dry matter is built up. By the time the corn plant reaches the silking phase, 40% of the total dry matter has been built up, and 60% of the total N has been taken in. This means that the 30 days between the V7 stage and the VT stage make up 40% of the total amount of N that the corn plant takes in.

N synchronization can be improved by not adding N until the plant is in the V7 stage. This won't damage the crop's yield. At 28 sites with different types of soil, the timing of when N fertilizer was added was the experimental variable. At the planting stage, the V7 stage, the V14 stage, and the silking stage, 180 kg N ha1 of ammonium nitrate was spread once. In most of the places where the N fertilizer was used, the corn yield went up. When the N application was put off until the V14 stage, only one of the 28 study sites saw a small drop in yield.

When N applications are delayed, it is possible that the weather could change the risk of yield loss. In many places, the highest yield was reached in dry years by not putting N on the surface of water-stressed corn until the V14 stage.

But the use of animal manure on many of the study sites, the use of soybeans as an early crop, and the use of different tillage systems across the sites have made this study more difficult. Two places where manure was spread on corn fields and then tilled will be included.

The type of crops that were grown before, how manure was managed, and how often the soil was turned over all affected how bad and when N deficiency happened. Contrary to what was said, there was a permanent loss of yield at one of the sites when N was added during or after the V6 stage.

This means that at the site, N availability must be sufficient before side dressing to ensure that the maximum yield is reached. As N deficiency went down with more time between side-dress N applications, the grain yield response to N went down. This shows that N deficiency levels were positively interacting with corn yield at the time of N application.

5.6. LEAF AREA INDEX

The ratio of the surface area of the leaves to the surface area of the ground is called the Leaf Area Index (LAI), and it shows how well plants can use light to make food. Some species and communities of plants have a direct relationship between LAI and their productivity.

For other species and communities, the relationship between productivity and LAI depends on factors like the canopy extinction coefficient, light, NUE, and the amount of light blocked by the canopy top. This is shown by the fact that C4 plants that grow in thick stands produce more NUE and leaf area than C3 plants that grow in the same environment.

Remote sensing has been used to come up with ways to figure out the LAI. Inversions of canopy radiative transfer models and empirical relationships between spectral vegetation indices and LAI. One problem with algorithms that are based on vegetation indices is that it is hard to apply their results to larger areas or different types of the canopy.

When measuring moderate to high levels of LAI, atmospheric and background effects, canopy architecture, solar target-sensor geometry, and lack of spectrum difference can make it hard to predict what the vegetation index will be.

Figure 5.3. AM300 device being used to measure the area, width, length, and perimeter of a leaf.

Source: Image by Wikimedia Commons

5.7. ENVIRONMENTAL INTERACTION

The main thing that affects how well crops grow is environmental stress. When the environment is bad, corn yields can drop by up to even more than 70%. Today's corn hybrids, which are made by breeding programs, have shown that they can handle environmental stresses and have more plants per area.

It's important to remember that only half of the yield increases in the modern era of breeding can be attributed to genetic improvements in corn. The other half is due to better ways of managing crops. When the soil moisture at a depth of 40 cm drops below 25%, the corn yield drops by a lot.

Adding water through irrigation, on the other hand, can double the yield. During the silking stage, when there is a drought, the ears can be empty. If there is a drought after the silking stage, the crop yield can drop by up to 20%.

Another study found that if there is moisture stress before silking, yield can drop by up to 25%. If there is moisture stress during silking, yield can drop by up to 50%. If soil moisture stress is still present after silking, the crop yield can drop by 21%.

Moisture stress can cause a lot of bad things to happen to corn plants, such as less grain yield, shorter cobs, less leaf area, and shorter stems. High temperatures are another thing that stresses out plants. At 45 degrees Celsius (113 degrees Fahrenheit), the rate of photosynthesis in corn can slow down by up to 95 percent.

Figure 5.4. Drought Stressed Corn.

Source: Image by Flickr

Stress on the corn from too much heat can delay the start of tassel growth. From a more moderate range of 22–17°C, high air temperatures of 32–27°C can reduce the rate of photosynthesis by 11 percent and the rate of total biomass production by 32 percent.

5.8. NITROGEN MANAGEMENT USING SITE-SPECIFIC TECHNOLOGIES

When data are collected using methods like satellite imagery, ground-based active-optical sensors, ground-based reflective sensors, leaf chlorophyll sensors, and aerial imagery or photography, there is no need to destroy a place or thing.

Remote sensing has been used in agriculture to figure out how land is used, how it is covered, and how much biomass is in crops. Now, remote sensing is used to figure out the status of crop N during the growing season.

A few studies have helped figure out the relationship between spectral reflectance, crop N status, and chlorophyll content.

Some of the first remote sensing methods used in studies were canopy reflectance/color photography, SPAD® (Konica-Minota Americans, Ramsey, NJ), and chlorophyll meters. Since the mid-1990s, the agricultural market and industry have been able to use a wide range of geospatial technologies.

Crop reflectance, color photography, and GBAO sensors have all been used to measure crop canopy spatial variability.

5.9. SOURCES OF NITROGEN AND INTERACTION WITH PLANT-SOIL SYSTEM

Theoretically, nitrogen can be found almost everywhere. Dinitrogen gas (N2) makes up about 78 percent (4000 trillion tons) of all the gas in the air. Organisms change dinitrogen into different forms, such as ammonium and nitrate.

About 95% of nitrogen is found in the natural organic matter of soil that hasn't been changed. But if you look at this element from an agricultural point of view, you'll see that crops and other plants get most of the nitrogen they need from certain sources. These sources include both those that come from nature and those that are made by humans.

Some of the natural ways that nitrogen gets into the soil are through mineralization and bacterial fixation. Ammonium and nitrate are the forms of nitrogen that plants can use from the soil. Nitrite, nitrous oxide, and nitrogen from the air are also found in soil. However, plants can't use these forms of nitrogen unless they are changed into ammonium or nitrate by bacteria or lightning.

The nitrogen that is stored in living things is another natural source. When living things die, they start to break down and put nitrogen back into the soil. Most plants and animals don't have as much nitrogen as legumes and dead animals, so they can both be used as natural fertilizers.

In crop rotations, legumes are used to fix nitrogen from the air and then deposit it in the top layers of agricultural soils. Animal bones or bone meal, which has a lot of nitrogen, can be used as a good alternative to chemical fertilizers. Chemically, NH3 (ammonia without water) is made from one part nitrogen and three parts hydrogen.

Anhydrous ammonia is the cheapest and most widely used form of nitrogen fertilizer because it is easy to use and easy to get. However, it is also a gas that is caustic, toxic, and explosive. So, there are strict rules about how this fertilizer can be used in the United States.

Urea has 46 percent nitrogen and is often used as a fertilizer in farming. Urea is easy to put into granules, so it is better for places where there is a lot of wind. In the soil, urea breaks down into CO2 and NH4+ ions. With the right care, urea is a great way to get nitrogen for crops.

The chemical formula for ammonium nitrate is NH4NO3. It is made up of nitrate and ammonium cation. Both ammonium and nitrate are forms of nitrogen that plants can use. Ammonium nitrate is just a combination of the two, and it is used as a stable high-nitrogen fertilizer, usually in places with a low pH and less wind.

Over 150 years ago, ammonium sulfate [(NH4)2 SO4] was one of the first nitrogen fertilizers made (IPNI, nutrient source specifics). Ammonium sulfate has N (21%) and S (24%) that are more concentrated and can be handled more easily. Ammonium sulfate is a great source of both nitrogen (N), and sulfur (S), which is very important to the way plants work, such as in making proteins.

Ammonium sulfate is the most popular fertilizer in Maine. It tends to lower pH and raise sulfur levels in soil, which makes it a better choice when crops also need more sulfur to grow and stay healthy.

5.9.1. What Is NUE, And How Can It Be Managed?

Nitrogen use efficiency (NUE) is a term that needs to be defined because it is not very specific. First of all, it is mostly an agricultural word. It has two main meanings, and when the term is used, both are meant at the same time.

The first way to define nitrogen use efficiency is by how well plants use and keep the nitrogen in the soil. This definition of NUE is used because plants are more likely to release nitrogen into the air as nitrous oxide after taking it in than to store it in their bodies. Leguminous crops have a high NUE because they take in nitrogen from the air and store it in their bodies instead of letting it escape into the air.

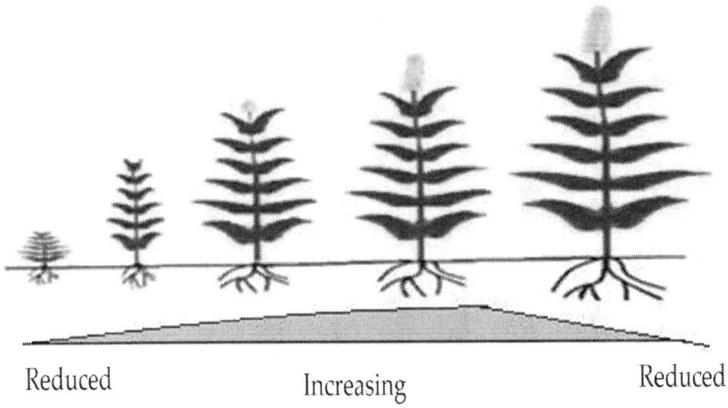

Figure 5.5. Plant N use efficiency at different growth stages.

Source: Image by Wikimedia Commons

There is still a lot of disagreement, though, about whether the net loss of nitrogen from legumes is less than that from other crops. Both types of crops lose some nitrogen when they break down. NUE measures how much nitrogen a plant takes in and how much of that nitrogen is lost through the plant's nitrous oxide emissions.

This shows how well plants use nitrogen and keep it. All of the nitrogen a plant takes in is thought to come from the soil, and all nitrous oxide emissions are thought to come from the soil before the nitrogen used to make nitrous oxide is taken up by a plant. This is because plants can take in and use both ammonium and nitrate forms of nitrogen through their roots.

This way of measuring NUE is the same as the first way nitrogen use efficiency is defined. The second definition of NUE is how well nitrogen added to soils by natural or artificial means is taken up by plants and not used for other things, like feeding anaerobic bacteria that cause denitrification or leaching by nitrogen dissolving in water.

Nitrogen getting into the water is often caused by too much fertilizer, too much soil drainage (which is often caused by a lack of organic matter), or growing plants at an angle. When there is too much water in fields, it leads to anaerobic denitrification.

NUE takes soil samples and analyses them in a lab to find out how much nitrogen was in the growing soil at the beginning of a season, how much of this nitrogen was fixed, and how much was lost through leaching and denitrification, and how much nitrogen is still in the soil. The nitrogen use efficiency of a system is higher if crops take in more nitrogen than is left in the soil or lost.

This is especially important when thinking about how to use fertilizers since the best thing would be for none of the fertilizer to go to waste and for all of it to be taken up by the plants it is used on. In general, nitrogen use efficiency (NUE) is the use of nitrogen in an effective way.

Nitrogen use efficiency is the amount of nitrogen a crop takes in and keeps until harvest compared to the amount of nitrogen that was actually available for the crop to use, with a focus on how much fertilizer is put on soils compared to how much nitrogen a crop takes in and keeps until harvest (Raun and Johnson, 1999).

Nitrogen use efficiency (NUE) has been used a lot as a way to measure the relationship between how much N is taken up and how much N is added. One way to explain NUE is to compare the amount of grain that was harvested to the amount of N that was added.

Because yield potential can vary, N can be lost from fields, and the prices of N fertilizer and corn can change over time, it's important to come up with fertilization practices that can make the best use of N fertilizer rates. NUE is a trait with many parts.

In conventional agricultural systems around the world, even in developed countries, nitrogen use efficiency is thought to be quite low on average. From just over 79 million pounds in 2002 to about 99 million pounds in 2012, nitrogen fertilizer was used a lot more around the world. But crop yields around the world haven't changed much in relation to how much nitrogen fertilizer has been used.

This shows how nitrogen fertilizer is used in a very inefficient way. Estimates show that about 33 percent of the nitrogen (N) in fertilizer is returned to the crop. This is a low NUE. Nitrate leaching or denitrification from too much rain is the main cause of N loss.

Farmers in Missouri who put N on their fields before planting say that putting N on in the fall is a big risk for N loss in the spring, which lowers crop yield. In the time between when N is added and when the crop starts to use it, N can be lost through leaching, clay fixation, immobilization, denitrification, and evaporation.

The low NUE of current N management practices is partly caused by the fact that N is not applied at the same time that crops need it. When you add a lot of nitrogen before you plant, it throws off the timing.

For example, the average amount of nitrogen used in the Corn Belt region of the United States over the past twenty years has been about 150 kg ha1, and 75 percent of those applications were made before planting (including the fall before planting).

During the first three weeks after emergence, corn uses less than 0.5 kg ha1 day1 of the mineral N from the soil. But after the first three weeks, the corn plant takes in more and more nitrogen until it has tassels. On average, it takes in 3.7 kg ha1 day1, and at its peak, it takes in 6 kg ha1 day1 (J.S. Schepers, personal communication).

Depending on the soil and the weather, pre-plant N could leach below the rooting zone of the crop early in the season, before the crop takes in the most N. So when a lot of N is added to the soil before planting, there will be a lot of available N in the soil profile before the plants actually take it up. This N could be lost over the next few weeks.

The effectiveness of a single N application before planting goes down as more N fertilizer is used. On the other hand, when N is added during the growing season instead of before planting, NUE goes up. NUE could go up if the crop got as much nitrogen as it needed.

5.9.2. Use of LAI, Soil Moisture, Sensor Timing, Relationship Equations, And Crop Height to Improve Optical Sensor's Crop Yield Prediction Models

Several studies were done to try to figure out how to improve the relationship between crop yield and sensor readings, but their results were all different, so it's not clear how useful they are. NDVI yield prediction models were also made better by using saturation-adjusted NDV and adding crop models to NDVI.

But because their results were not always the same, they were unable to give farmers an algorithm that could be used on a commercial scale. But LAI hasn't been used enough to give growers a stable and reliable sensor-based in-season N application equation.

It seems that the reason for this is that measuring LAI takes a lot of work. Sharma et al. did a study on potatoes that showed some promising results using estimated LAI from the sensor itself. This helped improve the relationship between the sensor and the yield of potatoes and also solved the problem of red wavelength saturation with more biomass.

This is an important finding of how to solve the red wavelength saturation problem and improve the relationship between the sensor and yield by using the same sensor to get NDVI readings. Using soil moisture to improve the relationship between sensor readings and crop yield was looked at in corn. A soil moisture meter was used in a research plot to do these studies.

In both studies, the relationship between crop yield and soil moisture got better when sensor readings were added. But farmers can't use soil moisture meter data on a large farm because it's not practical. So, rain data from a local weather station could help improve the relationship between sensors and crop yield. To show that it works, we need to do some research.

During the vegetative growth of corn, plant height alone has also been used as a metric. Hussain et al. (1999) say that the amount of water in the soil, the texture of the soil, the rate of fertilizer application, and the way the corn is grown all affect its height. High-resolution ultrasound distance sensing of the crop canopy can be used to measure the height of plants.

In Minnesota, the height of the sugar beet canopy (*Beta vulgaris L.*) was multiplied by a GS reading to get a better idea of leaf N concentration, sugar beet top N content, and dry matter yield. The NDVI measurement is linked to the leaf area index, which shows crop growth and development in two dimensions.

The relationship between the NDVI measurement and the leaf area index is best when corn is in its early to middle growth stages. As the corn grows older, the readings become more and more similar, so the relationship gets worse. Since the amount of nitrogen in sugar beet tops is related to both the leaf area and the leaf density, multiplying NDVI by the height of the canopy gives a "leaf volume" instead of a "leaf area index."

In experiments with other crop types with different leaf and growth structures, NDVI measurements were related to plant height in alfalfa (*Medicago sativa L.*), where plant height and surface area coverage develop almost at the same time, but not in grasses, where surface area coverage was almost continuous.

Using a sensor to measure height by hand is not practical in the real world with large fields. So, Sharma did a lot of research on both manual and automatic sensor heights for corn in order to improve the relationship between sensor height and corn yield.

They used an acoustic sensor to measure height through a sensor. This could be a way to make it easier to use crop height along with algorithms built with sensors, but the height sensor models need to be worked on more to make them more accurate.

Also, North Dakota State University was the first place to use the red edge wavelength and plant height along with the sensors to make nitrogen recommendations. An N calculator was made based on an economic return to help farmers use in-season sensor-based N recommendations for nitrogen before and during planting.

5.10. THE NEED TO USE PHOSPHORUS EFFICIENTLY

More and more people realize that there are limits to the world's rock phosphate reserves and that increasing the efficiency with which these reserves are used to grow crops is important if we want to keep or even increase agricultural productivity (Cordell et al., 2009).

The focus of this section is annual grain crops, which include cereals, oil seeds, and pulses. These crops provide 58 percent of the food energy for the world's growing population (FAO, 2010).

Improving the efficiency of phosphorus (P) fertilizer use for crop growth means making it easier for plants to get P from the soil (P-acquisition efficiency) and making it easier for plants to use P in ways that make them grow faster and send more biomass to the parts that can be harvested (internal P-use efficiency [PUE]).

Since only 15–30% of the P in fertilizer is taken up by plants in the same year, it is applied (Syers et al., 2008), improving the way plants take up P could lead to big efficiency gains.

This aspect of P efficiency has gotten a lot of attention and has recently been looked at (White & Hammond, 2008; Ramaekers et al., 2010; Wang et al., 2010; Richardson et al., 2011), which found promising opportunities for improved crop traits and agronomic measures.

On the other hand, physiological PUE hasn't gotten as much attention, and studies that compare PUE have often been confusing by differences in P-acquisition efficiency (Rose et al., 2011). Just trying to get plants to take in more P will help yields, but it will also cause more P to leave the field.

Increased exports of P can cause a lot of environmental problems away from the site (Tiessen, 2008; Childers et al., 2011), and they will need to be replaced with more fertilizer, in the long run, to keep P from being used up in the soil.

The most sustainable and productive agricultural systems will be those in which the amount of P exported is equal to the amount of P put in and which have high yields per unit of P used, or high PUE. In this section, we look at the characteristics of plants, from the whole plant down to the biochemical level, that help them make good use of P once they have it.

We also look at how efficient crop plants are in the bigger picture and look at ways to make them better.

5.10.1. P-Use Efficiency and P Dynamics in a Growing Crop

PUE is the amount of total biomass or yield that is made per unit of P taken up (Hammond et al., 2009). When necessary, the subscripts PUEt and PUEy are used to distinguish between PUEt and PUEy. When measuring biomass, most of the time, only the parts of plants that are above ground are taken into account.

In grain crops, PUEy is the amount of grain grown per unit of the plant's maximum height above ground. The productive use of a unit of P taken up by a plant depends on two things:

(A) How well it is used for metabolism and growth, and

(B) How long does it stay in the living parts of the plant where it helps with these processes.

Once a seed's P stores are gone, it needs to get P from the soil in order to keep growing. This time in a plant's life usually coincides with exponential growth, when there is a lot of soil P available because fertilizers have just been put down and roots don't have to compete for resources as much.

Even if soil water and nutrients are not used up, a crop's relative growth rate (RGR) will always go down due to self-shading, higher costs of breathing, and the death of older plant tissues and organs. At this stage, the roots continue to take in P, but the remobilization of P from dying tissues can become a very important source of P inside the plant.

At the time of reproductive growth, the main source of P for sink tissues is usually the remobilization of P from senescing vegetative tissues. This is because P uptake by roots decreases as root growth slows and soil P is used up.

In crops, a lot of the P in the plant parts moves to the grain, and the amount of P in the soil doesn't have much of an effect on grain yield at this stage (Romer & Schilling, 1986; Rose et al., 2008). The changes in the P

In this work, path analysis was used to figure out how much soil and leaf factors affected banana yield.

5.12.1. Soil Sampling and Analysis

At the beginning of July 2018, samples of soil were taken from the plow layer (0–20 cm). Fresh samples were taken from 10 centers in each plot and mixed well to make one sample that could be analyzed later. Soil samples were put through a sieve with a 2-mm mesh and split in half.

One part was kept at 4°c to find out how active the soil enzymes were. The other part was left out in the air to dry for chemical analysis of the soil.

Crowther and Large gave a way to compute soil urease activity (1956). With 10 ml of 10% (w/v) urea solution and 20 ml of citrate buffer, a 5 g sample of fresh soil was kept at 37°c for 24 hours (pH. 6.7). After that, 50 ml of 38°c water was added to the reaction mixture, and it was filtered.

The filtrate was diluted to 10 ml and then treated for 20 minutes at room temperature with 4 ml of sodium hydroxybenzene solution and 3 ml of sodium hypochlorite solution. At 578 nm, a part of an ultraviolet spectrometer was used to measure how much ammonium was released.

Urease activity was shown in terms of mg nh4-n per gram of soil per 24 hours. The 3,5-dinitrosalicylic acid (DNS) colorimetric method was used to measure the activity of invertase (Wang et al., 2009). 5 g of soil was mixed with 15 ml of 8% sucrose and 5 ml of phosphate buffer and left at 37°c for 24 hours (pH 5.5).

Figure 5.6. Soil samples are taken for further analysis. (Selian Agricultural Research Institute).

Source: Image by Flickr

One milliliter of the filtrate was mixed with three milliliters of DNS and put in a pot of boiling water for five minutes. The reaction solution was measured at 508 nm, and the activity of invertase was given as mg glucose per gram of soil per 24 hours. The disodium phenyl phosphate colorimetric method was used to measure acid phosphatase activity (Tabatabai and Bremner, 1969).

4 ml of 0.2 m modified universal buffer (pH. 6.5), and 1 ml of 25 mm p-nitrophenyl phosphate were mixed with 1 g of soil (PNP). After 1 hour of incubation at 37°c, 1 ml of 0.5 m CaCl2 and 4 ml of 0.5 m NaOH were added to stop the enzyme reaction. They were spun in a centrifuge and then filtered.

At 400 nm, spectrophotometry was used to measure absorbance. The enzyme activity was given as mg per gram of soil per 24 hours. Subbiah and Asija's method of hydrolyzing NaOH, diffusing NH3, and absorbing H3BO3 were used to find out how much soil N was available (1956).

Molybdenum blue colorimetry was used to measure the amount of soil P that was available (Olsen et al., 1954). The soil's available K was taken out with 1.0 m NH4OAc , and flame absorption spectrometry was used to measure it (Richards, 1993). The potassium dichromate oxidation method was used to find out how much was in the soil (Nelson and Sommers, 1982).

5.12.2. Leaf Sampling and Analysis

At the same time that soil samples were taken, leaf samples were also taken. The middle-upper part of the fully opened upper leaf on the main stem was taken as a sample. These samples were put right away into containers with liquid nitrogen for further testing. In each plot, a total of six plants were measured, and the mean value was found.

Vijayalakshmi et al. gave instructions on how to test the activity of nitrate reductase (2015). In short, 0.5 g of leaf tissue was added to 2.5 ml of potassium phosphate buffer with 5 mmol of cysteine hydrochloride, 1 mmol of EDTA, and 1 mmol of DTT. The 0.3 ml of supernatant was mixed with 0.1 mol potassium phosphate buffer, 0.1 mol potassium nitrate, and NADH solution.

By adding zinc acetate, the reaction was put to a stop. By adding sulphanilamide and N1-naphtylethylenediaminedihydrochloride, nitrite was measured at 540 nm. NR activity was shown in units of mol $NO2^{-}g^{-1}h^{-1}$ FW (fresh weight).

For the GS assay, 0.5 g of leaf tissue was extracted with 3 ml of Tris-buffer containing 0.765 g of Tris–HCl, 0.11 g of cysteine hydrochloride, 0.246 g of MgSO47H2O, and 0.02 g of EDTA. A Tris–HCl buffer, ATP, sodium glutamate, MgSO4, L-cysteine, and hydroxylamine were mixed with the enzyme extract. After letting the mixture sit for a while, 10% FeC13, 20% TCA, and 50% HCl were added.

A value of 540 nm was used to measure the absorbance of glutamyl hydroxamate. GS activity was measured in mol of -glutamyl hydroxamate per gram per hour (FW) (Vijayalakshmi et al., 2015). The SPS and SS activities were found by making small changes to the method described by Hu et al. (2015).

In short, leaf tissue was extracted with 100 mM Tris–HCl buffer, 5 mM MgC12, 2 mM EDTA-Na2, 5 mM DTT, and 2% polyvinylpyrrolidone. With 50 mM UDP-glucose, 50 mM fructose-6-P, 5 mM MgC12, 2 mM EDTA-Na2, and 5 mM DTT, the enzyme extract was incubated.

Using 100 L of 2 mM NaOH at 100°C for 10 minutes, the reaction was stopped. The solution was then let to cool, mixed with 1 mL of 0.1% resorcin and 3.5 mL of 30% HCl, and put in an oven at 80°C for 10 minutes. SPS activity was measured in units of g sucrose content per g per minute per FW.

The way SS activity was measured was similar to how SPS activity was measured, except fructose-dependent was used instead of fructose 6-P.

5.13. CONCLUSION

The chapter discusses improving nitrogen and phosphorus efficiencies for optimal plant growth and yield. This chapter also discussed the concept of soil and plant analysis. In this chapter, the use of tissue analysis for N management, leaf area index, and spatial variation has also been explained.

Towards the end of the chapter, the management of nitrogen with the help of site-specific technologies has been discussed. This chapter also explains the effect of nitrogen and phosphorus application on soil nitrogen morphological characteristics. In this chapter, the effect of different rates of nitrogen fertilizers on crop yield has also been discussed.

REFERENCES

1. Liu, D., Cui, Z., Yan, B., Gao, Y., Wu, B., Li, W. and Niu, J., 2020. Effect of nitrogen and phosphorus application on soil nitrogen morphological characteristics and grain yield of oil flax. *Oil Crop Science*, [online] 5(2), pp.29-35. Available at: <https://www.sciencedirect.com/science/article/pii/S2096242820300208> [Accessed 6 June 2022].
2. Sharma, L. and Bali, S., 2017. A Review of Methods to Improve Nitrogen Use Efficiency in Agriculture. *Sustainability*, [online] 10(2), p.51. Available at: <http://www.mdpi.com/journal/sustainability> [Accessed 6 June 2022].
3. Sharma, L., Zaeen, A., Bali, S. and Dwyer, J., 2018. Improving Nitrogen and Phosphorus Efficiency for Optimal Plant Growth and Yield. *New Visions in Plant Science*, [online] Available at: <https://www.intechopen.com/chapters/58114> [Accessed 6 June 2022].
4. Sun, J., Li, W., Li, C., Chang, W., Zhang, S., Zeng, Y., Zeng, C. and Peng, M., 2020. Effect of Different Rates of Nitrogen Fertilization on Crop Yield, Soil Properties and Leaf Physiological Attributes in Banana Under Subtropical Regions of China. *Frontiers in Plant Science*, [online] 11. Available at: <https://www.frontiersin.org/articles/10.3389/fpls.2020.613760/full> [Accessed 6 June 2022].
5. Veneklaas, E., Lambers, H., Bragg, J., Finnegan, P., Lovelock, C., Plaxton, W., Price, C., Scheible, W., Shane, M., White, P. and Raven, J., 2012. Opportunities for improving phosphorus-use efficiency in crop plants. *New Phytologist*, [online] 195(2), pp.306-320. Available at: <https://nph.onlinelibrary.wiley.com/doi/10.1111/j.1469-8137.2012.04190.x> [Accessed 6 June 2022].

CHAPTER 6

NANOTECHNOLOGY IN AGRICULTURE: NEW OPPORTUNITIES AND PERSPECTIVES

CONTENTS

The most pressing global concern in our world is ensuring food security for the world's largest fast-increasing population. According to projections, food consumption will climb from 59 to 98 percent when the world population approaches 9 billion by 2050.

But besides growth in world population, especially in developing countries, the world food supply has been disrupted by the use of bio-resources for energy production, manufacturing chemicals, high post-farming loss, less value addition, inefficient distribution and marketing systems, and other factors.

6.1. INTRODUCTION

Farmers all across the world will focus on employing new ideas and technology to increase crop yield through intense and extended agriculture. Present attempts are being strengthened by the use of nano-modified stimulants and precision farming.

Agricultural efficiency, soil development, securing water usage, food distribution in shops, and food quality are fundamental aspects of food security that may be enhanced by improvements in nanotechnology research.

Newer technology that increases productivity while decreasing food waste is critical to maintaining the nation's sustainable standard of living and improving food security. Nanotechnology can pave the way for thc production of high-quality meals in substantially enhanced practical forms, as well as the enhancement of nutritional bioavailability.

Many research studies are focused on expanding the use of nanotechnology in agricultural production and food processing. The growth of articles, intellectual property, and patents in nano-agriculture-based food, together with new research trends in food processing, nutraceutical distribution, packing, quality control, and serviceable food, is a rapidly increasing topic in nanotechnology research.

Nanotechnology is among the potential avenues for increasing the availability of food and developing innovative products for use in agriculture, food, water, the environment, medicine, energy, and electronics. It is a burgeoning and rapidly expanding region with novel and exclusive agricultural and food research applications.

Growing productivity and decreasing postharvest expenditure via improved outcomes with the assistance of newer scientific studies using nanotechnology and biotechnology in foodstuffs may be the ideal solution.

A few emerging topics for nanomaterials in agriculture include reducing the number of herbicides used, minimizing nutrient losses in fertilization, and increasing output through pest and nutrient control.

A few of the upcoming nanotechnology for food themes that can be greatly enhanced are smart nutrition delivery, bioseparation of proteins, a quick sampling of biological and chemical pollutants, nanoencapsulation of nutraceuticals, solubilization, and delivery.

Food nanotechnology entails the use of nanocarrier techniques to enhance bioactive substances' biological accessibility and barrier against various chemical or environmental variations. It improves sensory properties such as color, flavor, and texture, as well as food dependability. It also increased nutraceutical and medicine delivery system capture and biological convenience.

Nanotechnology is supporting the food industry by providing revolutionary food packaging solutions with improved mechanical, fencing, and antibacterial properties. Other advantages of the technique included nanotechnology-based sensors for trace detection, evaluating the status of diet in terms of transit and storage, and encapsulation of food modifiers or additional components.

Nanotechnology-based applications have increased the demand for nanoparticles in food biotechnology, science, food processing, food packaging, functional food creation, food safety, pathogen detection in food, and prolonged shelf-life of food and food supplies. Nanomaterials are particularly effective in improving food security and assisting the growth of the food production industry.

Depositing food manufacturing equipment (through biofilm coating), nanofabricated filters, sieves, and membranes, nanocomposite-based and nanosized adsorbents, and catalytic agents are further areas of use for providing aid in food processing. The use of nanoparticles in food packaging is intended to alleviate the difficulties encountered during material packing and to reduce the use of valuable raw materials and waste generation.

Nanotechnology, in general, provides the most promising opportunities for the development of new and enhanced goods. However, for many scientific groups, public concern about health dangers related to nanotechnology goods is being investigated further.

The purpose of this section is to describe the fundamentals of nanotechnology, its applications in agriculture, food research, postharvest, food process technology, packaging, and its role in supporting food security efforts, as well as the challenges associated with the technology's use in the food and agriculture system.

Agriculture, nano fertilizers, nano herbicides, nano pesticides, nanosensors, nanoformulations, gene sequencing or nano coding, nano coatings, nanoemulsions, nanocomposites, nanostructured materials, delivery, formulation, packaging, nanotubes, and nanotoxicity are the keywords.

Agricultural wastes, traditional inputs (land, water, energy, fertilizers, and pesticides), and a significant portion of plant protection products applied each year are lost or inaccessible to the target. Furthermore, agriculture (crop cultivation, livestock, and deforestation) is a major contributor to greenhouse gas emissions, accounting for approximately 24 percent of total yearly global emissions.

Another important concern in the primary sector is waste generation. Each year, European countries generate around 90 million tons of agricultural waste.

The European Commission has designated nanotechnology as one of six "Key Enabling Technologies" that contribute to long-term competitiveness and growth in a variety of industrial applications, therefore supporting the transition to a greener economy.

Agriculture remains a negligible area in comparison to other domains of nanotechnology use, such as health, materials, and energy. However, publications devoted to agricultural applications are increasing, as they are in other areas. This is illustrated by the growing number of peer-reviewed scholarly papers retrieved each year from the Elsevier Scopus database.

Last 10 October 2017, the query "Nanotechnology" and "Agriculture" was searched for. The results were confined to scientific publications, reviews, and conference papers published between 2000 and 2017. Scopus has 508 scientific products that have been indexed: 264 (52%) articles, 143 (285%) reviews, and 91 (18%) conference papers were submitted

In terms of publication distribution among the most productive nations, the United States and India share the research leadership in this sector, having published around 63 percent of articles jointly (27 and 25 percent, respectively). China ranks third (10%), with the EU nations accounting for almost 20% of all articles.

Nanomaterials' distinctive physicochemical qualities, such as catalytic reactivity, large surface area, size, and form, have the ability to open new paradigms and bring new agricultural tactics. Such new paradigms necessitate the development of new terminology.

The word "Agri-nano techniques" was used in a newly released book. Because no particular definition for this term was supplied, it has been used to refer to nanosystems employed for nutrient delivery in crops.

The word "Phyto nanotechnology" was coined to describe the application of nanotechnology in plant production systems or, more generally, in plant science. Yet, because nanotechnology applications in agriculture are still in their infancy, it is extremely conceivable that new terminology will be coined in the future to describe more precise technical breakthroughs.

6.2. FUNDAMENTALS OF NANOTECHNOLOGY

Nanotechnology is defined as the "synthesis, design, characterization, and application of assemblies, tools, and systems by influencing the morphology and size variation at the nanoscale level from 1 to 100 nm." For your convenience, one nanometer equals one billionth (109) of a meter, implying that the technique can be applied at this scale.

Nanoscience and nanotechnologies are novel attitudes in the developmental study associated with the discovery of wonders and the operation of substances at the atomic, molecular, or macromolecular levels, at which point their properties differ significantly from those at the bulk level.

The biological, chemical and physical properties of the resultant products differ significantly from those of the bulk material. The study of qualities at the nanoscale yields changed properties that may be utilized to create innovative materials with modified structures, newer tools, and goods that perform better.

The nanoscale features of materials supplied newer properties such as increased strength, better optical properties, changed antibacterial potential, and exceptional superconductive nature. They can take on the forms of nanotubes, nanoparticles, nanofibers, fullerenes, nanosheets, and nanowhiskers.

A nanomaterial is described as a "natural, incidental, or industrial material having particles, in an unbound state or in the shape of aggregation or agglomeration, where 50% or more of the particles in the number and size distribution, one or more than one dimension resides in the range of 1–100 nm".

6.2.1. Approaches of Nanotechnology

Even in food technology, nanotechnology may be deployed using two opposing approaches: "bottom-up" or "top-down." The physical technique used for food and agriculturally based commodities may be used to utilize the top-down strategy.

Commercial nanomaterial production now relies mostly on the "top-down" strategy, in which nanoscale materials are manufactured through size reduction of bulk precursors, milling technology, nanolithography, or precision engineering. The dry milling process is used to obtain grain flour with smaller particle size and hence better water-retaining capacity.

Figure 6.1. Nano-lithography on pristine oxide surface.

Source: Image by Flickr

Through a size reduction method, the top-down strategy can improve the antioxidant qualities of green tea. Based on the report, powdered green tea with a particle size of 1000 nm has greater nutrient digestion, which leads to an increased ability for dismutase enzyme with increased oxygen elimination and hence boosted antioxidant activity.

Another top-down strategy is homogenization, which is often employed in the dairy industry for size reduction in the case of globules, vaporization, and laser application linked with chilling. A surface area with superior qualities is the functioning of food material for the needed purpose. Top-down and bottom-up techniques are demonstrated and are dependent on size reduction since finer size material has higher porosity.

The top-down technique is related to nanotechnology-based devices that are monitored with external power to yield the necessary parameters and particular initiation from greater dimensions requiring cutting to reach the appropriate size. Via sophisticated processes, atoms are turned into nanoscale materials or fixed self-assemblies using the bottom-up technique.

Figure 6.2. Liliana Stan of the Center for Nanoscale Materials (CNM).

Source: Image by Flickr

Matching attracting and repulsive interactions between molecule pairs used as building blocks for effective supramolecular structures is required for self-assembly. Self-assemblage or clustering of casein micelles, for example, results in the production of carbohydrate-binding proteins.

Casein micelle structures, protein-polysaccharide liposome complexes, and aggregation are all examples of self-assembled nanostructures in food. As shown in research, the bottom-up strategy allows for the creation of nanostructures with fewer defects, improved identical chemical systems, and better organization.

6.3. THE ROLE OF NANOTECHNOLOGY IN ACHIEVING FOOD SECURITY

Nanoparticles are utilized in the food processing industry to improve food durability and color retention. Silicate nanoparticles can reduce oxygen flow and moisture loss in packing containers. This keeps the food fresh for a longer period of time. Several nanoparticles can also selectively bind to infections, removing them entirely in the process.

Biofilm production can be inhibited by nanoparticles. Biofilms are densely packed bacterial cells that adhere to a variety of substrates and form a barrier that prevents any type of penetration. These biofilms cause issues in the food processing industries, such as biofouling, biocorrosion, and accumulation.

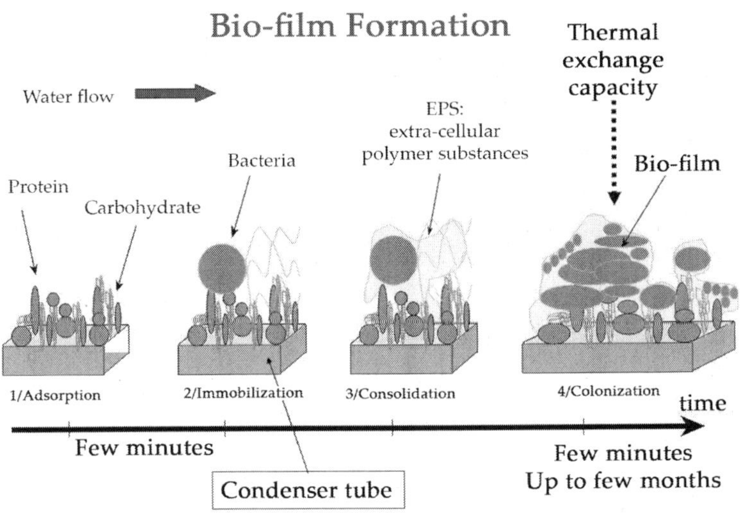

Figure 6.3. Biofilm formation.

Source: Image by Wikimedia Commons

Glycerol monolaurate is an antibacterial substance that inhibits the growth of various Gram-positive bacteria. This is used to prevent the production of biofilms by three distinct strains of *Staphylococcus aureus* and MRSA. Organic substances (essential oils) are antibacterial and very sensitive to harsh physical conditions.

Inorganic nanoparticles, on the other hand, exhibit high antibacterial action at low concentrations and are more resilient in harsh circumstances.

As a result, manufacturers have lately expressed a keen interest in using these nanoparticles in antimicrobial food packaging.

Due to rising climate change, land limits, expanding population, industrialization, low productivity, and substantial postharvest losses, the globe is facing numerous and unprecedented difficulties. This has resulted in global food insecurity.

Around 2 billion people worldwide suffer from moderate to severe food insecurity. They are more vulnerable to hunger and bad health since they do not have regular access to food. This dramatically changed society necessitates new perspectives on food insecurity and its effects.

To achieve the needed food on a global scale, Scientists from all around the world have proposed many strategies and procedures for increasing agricultural yield, reducing postharvest losses, and assuring sustainability.

Providing a healthy food supply to protect people's health and well-being requires innovative technologies. Recent research has demonstrated the potential of nanotechnology to improve agriculture by increasing the efficiency of agricultural inputs and bringing answers to agricultural challenges to improve food output and security.

The following section discusses the role of nanotechnology in agriculture, postharvest loss reduction, and food processing in ensuring food security.

6.3.1. Agriculture

Agriculture is the growing of various crops and the breeding of livestock for the production of food. It is regarded as the backbone economy for the majority of emerging countries, playing an important role in growth and development. As the world's population grows, so does the demand for more food, and scientists and engineers are developing innovative techniques to improve agricultural productivity.

Agriculture nanotechnology has recently concentrated on research and application to address agriculture and environmental challenges such as sustainability, crop enhancement, and increased production.

Agricultural nanotechnology appears to be quite attractive for poor nations in terms of reducing hunger, underfeeding, and child mortality. Developed and growing countries such as Germany, the United States of America (USA), Brazil, China, India, France, and Korea have expanded their interest in employing nanomaterials for agricultural purposes, as evidenced by a rise in the number of articles and patents.

Nanotechnology, as a prospective gadget, may be used to remodel agricultural divisions; it aids in studying the biochemical routes of crops by updating conservative approaches for assessing environmental difficulties and their application to production improvements.

Comparisons of nanotechnology with environmentally friendly technologies and agricultural biotechnology reveal the possibility of greater and faster influence on all elements of the agro-value nexus for synchronized public benefits and legal, moral, and environmental repercussions.

The potential use of nanoscale agrochemicals in agriculture, such as nano fertilizers, nano pesticides, Nanosensors, and nanoformulations, has revolutionized traditional agricultural techniques, making them more sustainable and efficient.

There are several uses of nanotechnology in agriculture, including wastewater treatment, improving the quality of contaminated soil, and increasing agricultural output via security in the form of sensors to detect infections.

Nano biosensors, for example, are broad-ranging nanotools that support the establishment of high-tech agricultural farms while also demonstrating the practical and planned uses of nanotools in terms of agricultural input control and management accuracy.

One good influence of nanotechnology in agriculture is the use of nanopore carrying zeolite for delayed discharge and better efficacy of enrichers, Nanosensors for detecting soil quality, and smooth supply systems for herbicides. Carbon, silica, silver, and alumina-silicate nanoparticles are among the nanoparticles used to detect plant diseases.

The use of such nanomaterials in agriculture is recommended to eliminate chemical spraying by ensuring a steady supply of energetic molecules. It can reduce nutrient waste during fertilizer application and increase harvests by improving water and ingredient management.

The responses of many rice cultivars exposed to engineered nanoparticles at various phases of development and under various settings were also reported.

6.3.2. Nanotechnology to Control Plant Diseases

Every year, plant pests and diseases destroy 20–40% of the world's crops. Pest control in current farming techniques is mainly reliant on the use of pesticides such as insecticides, fungicides, and herbicides. It is critical to produce low-cost, high-performance insecticides that are less hazardous to the environment.

Nanotechnology, for example, can provide benefits to pesticides by lowering toxicity, extending shelf-life, and enhancing the solubility of poorly water-soluble pesticides, all of which might have a favorable environmental impact.

The importance of agricultural nanotechnology, particularly for disease control and safety, has been described elsewhere Nano-based traditional herbicides and pesticides aid in the gradual and consistent delivery of nutrients and agricultural chemicals to plants in a regulated proportion.

Nanoparticles might possibly play an important role in the management of insect pests and host infections. For the manufacture of nano-insecticides, polysaccharides such as chitosan, alginates, starch, and polyesters have been investigated.

In general, nanoparticles can be used to protect plants through two mechanisms:

(A) Nanoparticles offering crop protection, or

(B) Nanoparticles act as carriers for current pesticides and can be delivered by spray.

However, the application of nanoparticles in plant protection and food production remains underutilized.

6.3.3. Nanotechnology to Improve Quality of Soil and Fertilizer Distribution

Agricultural management nanotechnology is a critical technology for increasing crop output. Carbon nanotubes, nanofibers, and quantum dots, for example, are currently being used as biosensors in agriculture research to assess soil quality and fertilizer distribution.

The goal of nanoparticles is to prevent chemical dissemination, reduce nutrient loss during fertilization, and boost quality and production with optimum nutrition. The creation and application of vermiculite, nano clay, and zeolite might boost fertilizer efficacy and crop output in coarse-textured soils for ecological agriculture.

In ecological agricultural systems, amending sandy loam soils with inorganic amendments reduces NH4–N transit while improving N fertilizer production. On the basis of chemical composition and nanoparticle form, nano clay is classified as montmorillonite, bentonite, kaolinite, hectorite, and halloysite.

The majority of agricultural output is primarily reliant on fertilizer use. According to research, crop yield is linearly controlled by the use of fertilizers to boost soil fertility. The application of nano fertilizer is critical for increasing crop yield.

Nano fertilizer is a nanometer-sized substance that increases nutrient delivery to plants and manages the progressive release of nutrients into the soil in a carefully regulated manner, hence preventing eutrophication and water pollution.

The use of nanostructured or nanoparticles for fertilizer transport or limited release pathways to construct smooth fertilizer as new options to adjust nutrient utilization efficacy and lower charges for environmental safety is enabled by nanotechnology.

Nano-fertilizers might enhance nutrient efficiency by encapsulating nutrients into nanoparticles, which can be accomplished in three ways.

(A) Nutrient encapsulation inside nanoporous structures,

(B) Thin polymeric film coating, or

(C) Delivery in the form of nanoscale particles or solutions. Because of their tiny size, nanoscale fertilizers may be able to reach plant surfaces and transport pathways, resulting in more effective nutrient delivery.

Nano-fertilizer derived from banana peels was utilized to promote the development of tomatoes, peppers, and flowers. Nano fertilizers were employed for crop development and enhancement, such as nanoparticles of ZnO for chickpea, silicon dioxide and iron slag powder for maize, colloidal silica, and NPK for tomato, TiO2 for spinach, and gold and sulfur fertilizers.

Fertilizers may be treated with nanoscale transporters in such a manner that they link the roots of the plant with the surrounding soil contents and organic material, reducing chemical loss and environmental concerns.

Nanoscale fertilizers can lower soil toxicity, reducing the possible negative effects associated with high doses. Such nano fertilizers limit the release of nutrients and lengthen the duration of fertilizer action.

TiO2 nanoparticles have had a significant impact on maize crop development; moreover, SiO2 with TiO2 nanoparticles boosted the action of nitrate and increased plant absorption potential through the effective use of water and fertilizer.

6.3.4. Nanotechnology and Gene Sequencing

Nanotechnology also aided gene sequencing, which increased the identification and utilization of characteristic plant mechanisms, improving their ability to adapt to environmental stressors and diseases. Quantum dots and nanoparticles have been shown to be highly accurate biological markers. Optical DNA mapping in the age of nanotechnology and nanoscopic has also been described.

Optical DNA mapping allows researchers to acquire genetic and epigenetic information on individual DNA molecules. Nanopore sequencing might enable the sequencing of single DNA molecules spanning tens of kbp (perhaps up to 100 kbp), overcoming the constraints of short-read data. Nanotechnologies have been used in a wide range of applications.

Genome sequencing, targeted resequencing and transcription factor binding site finding, noncoding RNA expression monitoring, and molecular diagnostic.

Farming, as well as the food industry, are a significant source of income and employment for a large portion of the people. Agriculture has a critical part in self-sustaining economic growth by providing fundamental ingredients to humanity as well as raw materials for industrialization. Agriculture employs over 2.5 billion people worldwide, according to estimates.

Agriculture is a diverse area that is seeing rapid technological advancement. Many modern agricultural technologies are being launched to boost production by minimizing the direct and indirect variables that impact crop productivity.

Insects, diseases, and weeds are major yield loss issues that can be addressed by humans using insecticides, fungicides, and herbicides; accordingly, to regulate these biotic variables and to boost output, high doses of chemical pesticides and fertilizers are administered to growing crops.

As a result, residue pesticides and chemicals are polluting soil, food, and water. When tainted food is ingested, it causes a slew of major problems for the customer. Aside from chemicals, various bacterial infections and toxins generated by microorganisms are important food pollutants, causing further health concerns.

One excellent method for preventing contaminated food-related health concerns is to identify food contaminants, whether chemical or biological, before ingestion.

This will assure food safety. There are several traditional and innovative methods for detecting biological and chemical contaminants in food samples, such as culture plate technique, chromatography, spectroscopy, immunology, molecular biology technique, etc.; however, they are all time-consuming, costly, and have limited sensitivity.

Agriculture is the primary source of food for humans and cattle. A variety of biotic and abiotic elements are posing challenges to agricultural output and productivity. To ensure food security for the world's rapidly rising population, food output and productivity must be massively increased.

There is a need for advanced technologies to maximize output while maintaining excellent quality assurance. Identification, diagnosis, and avoidance of risks in order to attain regional and global food security. Thus, quick, real-time, portable, and cost-effective solutions in the agricultural and food sectors are desired to improve customer livelihood and resource use.

Many technologies have recently been invented and changed the agricultural industry, with nano technology being the most promising. Even though its practical application is minimal at the moment, it has a lot of potential in the near future to improve agricultural practices over conventional farming at various stages from crop production to postharvest, thus thriving the agriculture sector by increasing food production and crop productivity.

Crop productivity or yield increase is often attainable in two ways:

1. Besides reducing yield loss induced by multiple crop phases such as insect and disease (Biotic stress), numerous adverse environmental factors (Abiotic stress) like water stress, high-temperature stress, salt stress, cold stress, harmful radiation, and nutrient deficiencies all through agricultural production, and attempting to avoid losses after harvesting farm product (Postharvest stages)
2. Improving yield by implementing significantly enhanced advanced crop production systems, hence lowering input costs and boosting output with a high cost-benefit ratio.

Nanotechnology may be used in both techniques to increase agricultural and food output and productivity. Nano biosensors, or biosensors made of nanomaterials, are one of the most important applications of nanotechnology.

They are created by combining various departments such as bioelectronics, material science, miniaturization techniques, electrode design, fabrication technology, nanolithography, and microfluidics. A biosensor is self-contained integrated equipment that detects and characterizes biological molecules. Improvements in the fundamental properties of biosensors will lead to widespread application in critical sectors of food and agriculture.

In this study, the current function and uses of nano biosensors in the agricultural and food industries are studied, as well as the potential of nano biosensors, probable applications with rapid development, and advantages in the future.

6.4. BIOSENSOR

A sensor is a device, module, machine, or subsystem that detects events or changes in its environment, such as temperature, humidity, water flow, light intensity, and so on, and sends the information to another electronic circuit or electrical instrument for measurement and/or analysis.

A biosensor is a sort of sensing technology that consists of a biological component, such as a cell, enzyme, or antibody, that is coupled to a small transducer, which is a device that converts one type of signal into another so that it may be readily detected by another system.

Biosensors enable the identification of changes in cells and molecules even at extremely low concentrations of the tested substance. When the drug binds to the biological component, the transducer generates a signal proportional to the amount of substance present.

For example, if a food contains a higher concentration of germs, the biosensor will provide a strong signal suggesting that the item is dangerous to consume. With this technique, large quantities of food may be easily verified for safety before consumption.

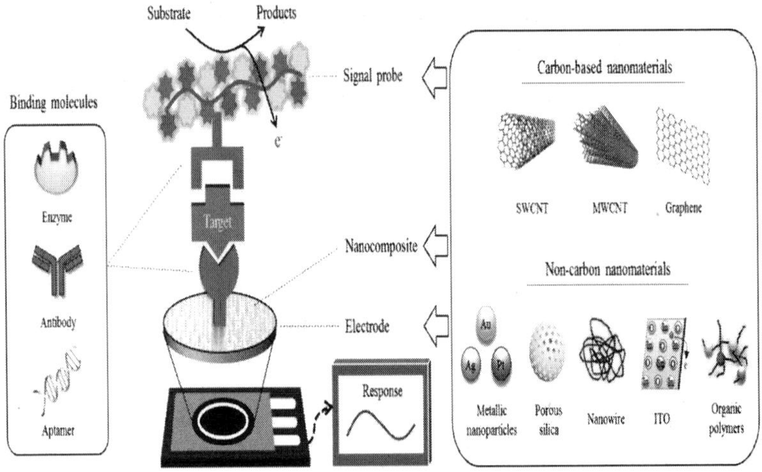

Figure 6.4. An analytical principle scheme for electrochemical biosensors based on carbon and non-carbon nanomaterials.

Source: Image by Wikimedia Commons

6.5. NANOBIOSENSOR

Nano biosensors are biosensors that are created utilizing nanomaterials. This sort of biosensor can accurately detect at the atomic level with greater precision. Normal biosensors have receptor and transducer materials that are micro and macro in size; however, nano biosensors have a receptor or a transducer that is nano in size, i.e., at least one dimension is smaller than 100 nm.

Because nanomaterials are so tiny, they have distinct physical, mechanical, optical, electrical, and magnetic properties as compared to ordinary materials. This is the extra benefit of the nanosensor and the rationale for its improved sensing abilities. Researchers employed a variety of nanomaterials and nanocomposites to improve sensitivity, shelf life, and precision in biosensing findings.

For diverse field applications, four types of nanomaterials are commonly used: carbon-based, metal-based, polymer-based, and nanocomposites. Dendrimer, among the four, is not employed as frequently as other nanomaterials in the biosensing sector.

To increase electrochemical characteristics, carbon nanotubes (CNTs), quantum dots (QDs), gold (Au), silica, silver (Ag), graphene, and other nanocomposites are created with a high surface area to volume ratio.

Molecular bonding is a branch of biological surface science that is closely connected to the study of how to modify the characteristics of nanostructures by regulating their structure and surface at the nanoscale level. Recent advances in nanotechnology allow for the synthesis of nanomaterials with remarkable optical and electrical characteristics for electronics and sensing applications.

The efficiency of biosensors may be increased by increasing the detection limit and overall performance by employing nanomaterials. Nanomaterials provide a favorable environment for the assembly of bio-recognition elements, as well as a high surface area and electrical conductivity that raises the detection limit.

6.5.1. Nanobiosensor Application in Food

The most prevalent sources of food contamination include harmful microorganisms such as pathogenic bacteria, viruses, or parasites and their toxin, as well as excessive use of agrochemicals (pesticides, herbicides, and food preservatives).

Pathogenic bacteria such as *Bacillus cereus, Clostridium botulinum, Escherichia coli (E. coli), Listeria monocytogenes, Salmonella,* and *Staphylococcus aureus* are the major contaminants that cause foodborne disease. Currently, there are few ways to detect toxins and foodborne pathogens in the field, making early detection of probable contamination challenging due to inefficient traditional approaches.

The use of biosensors with different nanostructures such as thin films, nanorods, nanoparticles, and nanofibers in analytical methods for detecting food pollutants has enhanced detection sensitivity and mobility.

Nanosensors or nano biosensors are used in food microbiology for pathogen detection in processing facilities or food material, measurement of accessible food ingredients, and notifying consumers and distributors of the safety status of food.

6.5.2. Nanobiosensors for Food Pathogen Detection

The traditional approach for detecting food pathogens is colony counting (CFU) on an agar plate, which takes 2–3 days for preliminary results and approximately 1 week to confirm pathogen specificity. These traditional procedures are unsuitable for very perishable foods.

Traditional CFU techniques can be replaced by polymerase chain reaction (PCR) and enzyme-linked immunosorbent test (ELISA). However, these procedures are time-consuming and expensive. So, utilizing nano biosensors that can be adapted to portable platforms, we can enable quick testing of a wide range of pathogens with the possibility of on-site analysis.

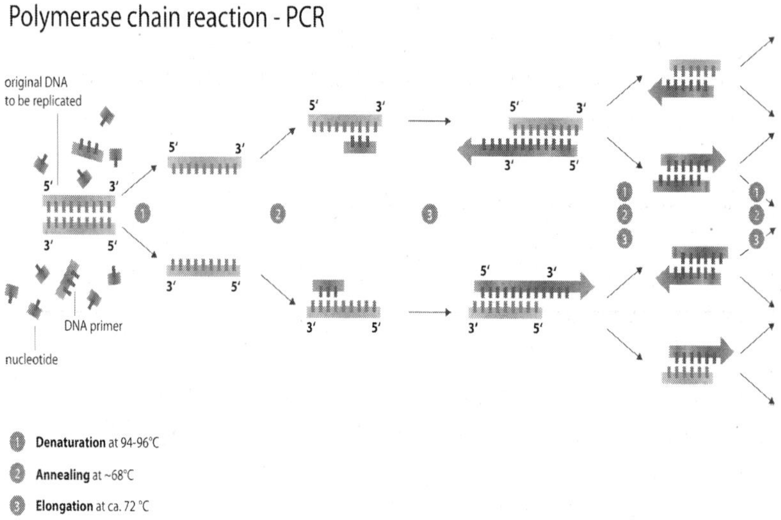

Figure 6.5. Different ELISA variants: direct ELISA, indirect ELISA, direct sandwich ELISA, indirect sandwich ELISA.

Source: Image by Wikimedia Commons

With just an electrochemical impedance spectrum, a dimethyl siloxane microfluidic immunosensor integrated with a specific antibody immobilized on a nanoporous alumina membrane was developed for rapid detection of foodborne pathogens *Escherichia coli* O157:H7 and *Staphylococcus aureus*.

Carbon nanotubes have been employed to construct biosensors for the detection of foodborne pathogenic bacteria (*Staphylococcus aureus*) in fresh meat due to their high electrical conductivity and abundance of functional groups on their surface.

Miranda et al. created a hybrid colorimetric enzymatic nanocomposite biosensor based on enzyme amplification for detecting *E. coli* in aqueous solutions. The method's effectiveness was shown in both solution and test strip formats. β -galactosidase, an anionic enzyme, is electrostatically linked to cationic Gold nanoparticles (AuNPs) with quaternary amine head groups, inhibiting enzyme activity.

As AuNPs bind to bacteria, the connected β-galactosidase is freed, regaining its function, and the binding activity and color creation as a result of the enzymatic process were assessed using colorimetric methods. Bacteria may be identified with this approach at quantities of 1 x 102bacteria/mL in solution.

Escherichia coli (*E. coli*) O157:H7 has been discovered utilizing a generated beautiful 3D architecture of silver (Ag) nanoparticles as nanoflowers as the interface material for electrochemical biosensing in an effort to assure food safety.

For effective conjugation of Ag nanoflowers, bovine serum albumin (BSA) was utilized as a stabilizing agent. *Salmonella spp.* is a pathogenic food bacterium that causes salmonellosis disorders. Millions of individuals worldwide are infected by this disease each year. Salmonella in milk is detected using magnetic nanoparticles and TiO2 nanocrystals.

The pathogen is initially trapped by antibody-immobilized magnetic NPs in this approach. The MNP–Salmonella complexes are then bound by antibody-conjugated TiO2, as measured by absorbance. This approach is sensitive enough to identify Salmonella in milk samples at 100 CFU/ml.

6.5.3. Nanobiosensor in Crop Stress Management

Crop growth and output are affected by biotic, abiotic, and nutritional stress. Plant hormones, also known as phytohormones, play a significant part in the control and regulation of physiological processes of development, as well as in stress response and regulation in plants.

Abiotic and biotic stress causes plants to produce a variety of unique chemical metabolites and plant hormones in order to prepare for harsh conditions in the surrounding environment. In terms of detection, Nano biosensors have played a critical role in resolving problems connected to phytohormone imbalance and its implications.

So that subsequent plant recovery activity may be implemented as soon as possible, it prevents significant plant damage and production decline.

Nano biosensors have made a substantial contribution to the long-standing objective of precision in agriculture.

A label-free electrochemical impedance immunosensor for abscisic acid detection has been created employing an anti-ABA antibody adsorbed on a porous nanogold surface. To get the required outcome, multiple matrices were tested using the same biosensing approach.

The results showed that abscisic acid in plants might be detected using an antibody-based nano biosensor. Electrochemical impedance spectroscopy using a successfully produced porous nanogold sheet and subsequent modification of the glassy carbon electrode detects gibberellic acid (GA) in hybrid rice grain samples.

For the measurement of salicylic acid in plasma and milk samples, a simple amperometric biosensor made of graphite covered with polypyrrole (PPy) was created. A real-time, highly selective nano biosensor for measuring cytokinin and auxin concentrations in tomato xylem sap exudates has been created.

In the event of plant stressors, it is advised that measures be taken rather than cures or treatments since this will result in massive crop loss and low-quality food at the disaster stage. At the moment, it has been discovered that the biosensor's operational stability is restricting its technical implications in the agricultural industry.

However, with the complete development of nano biosensing technology, this early detection of worrisome stress situations (biotic and abiotic stress) in plants will be attainable in the future. Certain Nano biosensors have been created in this area to detect indicators or signaling chemicals of plant stress situations.

A fluorescence-based biosensor that only gives data regarding hormone distribution under plant abiotic and biotic stressors was recently created to study jasmonate signaling in plants. Water is essential not just for photosynthesis but also for the movement of nutrients and numerous microelements required for crop health.

An optical fluorescence biosensor for detecting plant water stress has been disclosed. Plant development, physical strength, grain formation, and yield will be hampered if macro and micronutrients are deficient. The detection of inadequate nutrients using nanosensors will aid in the improvement of plant development and productivity.

Enzyme-based phosphate biosensors with fluorescent nanoparticles that uses a fluorescence transduction mechanism to detect phosphate and other nutrients, including nitrogen, calcium, and zinc.

6.5.4. Nanobiosensors in Smart Farming

Nanosensors are extremely small and can detect soil conditions, irrigation requirements, pH of water and soil, nutrient requirements, disease and insect incidence, soil temperature, and other vital parameters by dispersing them across the field.

Need-based action would boost agricultural yields while reducing unneeded personnel resources such as fertilizer, pesticides, and so on, based on the parameters collected by the nanosensor. This nanosensor idea is perfectly suited to the objective of precision farming or smart farming.

Nanosensors constructed of non-biological materials, such as carbon nanotubes, have the capacity to sense and communicate, serving as wireless nanoantenna due to their tiny dimensions, and may gather data from a variety of sources. External devices can then use the data to provide extremely precise reports and adapt to potentially disastrous changes in their surroundings.

For example, networked nanosensors for monitoring soil or plant conditions can notify automatically based on circumstances identified by sensors, influencing the more effective use of fertilizers, herbicides, pesticides, insecticides, and so on.

For detecting plant stress indicators, Nano biosensors are presently being created with all integrated elements such as a power supply, sensing unit, detector, and display unit in a single chip.

Various suggestive signals, such as an increase in sucrose content, a change in the concentration of nutrients and hormones, and so on, can be utilized to further translate them into visual indications using biosensing technology of signal processing.

6.6. NANOTUBES IN FOOD AND AGRICULTURE

Buckyballs with two sides attached with different atomic groups included in the conventional hexagonal shape to form a hollow carbon tube are known as nanotubes.

Nanotubes can tolerate high temperatures and have a flexible and robust character, making them useful in medical instruments, alumina, sports equipment, and food processing equipment. Inorganics, carbon, bio-microtubules, viral proteins, amyloid proteins, porins, lactalbumin, DNA, carbohydrates, lipids, synthetic polymers, and other organics are used to create nanotubular textures.

The synthesis and use of enzyme-based nanotubes as enzyme-supporting substrates have been shown to improve the catalytic activity of nanotube-deposited lipases.

Carbon nanotubes increase the mechanical qualities of food packaging. Carbon nanotubes and polyamides have been used to boost the tensile strength of various polymers. Carbon nanotube-based biosensors have been used to detect bacteria, hazardous components, and other metabolites in drinks and food due to their quicker detection, easy technique, and lower cost.

TiO2-based carbon nanotubes have been shown to have increased disinfection potential against *Bacillus cereus* spores. Silver-deposited TiO2 nanoparticles enhanced their bacterial killing capabilities against *E. coli*. Carbon nanotubes' strong antibacterial properties caused them to be destroyed by direct touch.

Partial hydrolysis of α-lactalbumin (milk proteins) resulted in the formation of nanotubes. Self-assembly of nanotubes from partly hydrolyzed α-lactalbumin. According to a research article, α-lactalbumin nanotubes have improved viscosity and hardness and can therefore be used as a thickening.

According to the study, -lactalbumin nanotubes are useful in pharmaceutical and food applications as viscosities, gelation agents (sol-gel formation) with regulated assembly, encapsulation agents with fixed taste, and drug delivery, among other things. Nanotubes also have been related to contemporary agricultural operations.

Various nanoparticles, according to some findings, can infiltrate plant cell walls. It was previously shown that carbon nanotubes might penetrate tomato seeds and influence their sprouting and development. Germination and development of tomatoes treated with various concentrations of carbon nanotubes (0, 10, and 40 g/L) demonstrated a remarkable increase in biomass.

This is attributed to improved water intake by carbon nanotubes, which improves the plant. Mesoporous silica nanoparticles produced from gold also pierced the cell wall and aided in the entry of DNA into the cell.

6.7. NEED FOR NANOSENSORS

Food packaging and transportation are the areas of food production where the demand for Nanosensors is most obvious.

Consumers must rely on expiry dates supplied by manufacturers based on a set of idealized assumptions about how the food is stored or transported because food packaging prevents sensory exposure from the goods. If shipping or storage requirements are broken for any length of time, the quality of food may decrease, which the consumer may not be aware of till the food item is opened or eaten.

Because of their particular chemical and electro-optical features, Nanosensors can mitigate the drawbacks of food packaging. They can detect gases, odors, chemical pollutants, infections, and even changes in environmental conditions. Nanosensors ensure that customers buy fresh and appetizing items while also lowering the incidence of foodborne diseases, and improving food safety.

6.7.1. Nanosensors in Food and Agriculture

Nanosensors are arranged similarly to conventional sensors; however, they are manufactured at the nanoscale.

As a result, a nanosensor may be characterized as a very small device that can bond to whatever is being detected and transmit back a signal. These small sensors can detect and respond to physicochemical (sensors) and biological signals (biosensors), converting the reaction into the signal or output that people can utilize.

When compared to traditional sensors and their drawbacks, Nanosensors offer various advantages, such as high sensitivity and selectivity, near real-time detection, cheap cost and mobility, and other important features that are enhanced by the use of nanomaterials in their creation. There are several methodologies for developing Nanosensors, including top-down lithography, molecular self-assembly, and bottom-up assembly methods.

Current Nanosensors devices are classified as follows: Nanostructured materials, such as porous silicon, Nanoparticle-based sensors, Nanoprobes, Nanowire Nanosensors, Nanosystems: cantilevers, Nano-electromechanical systems (NEMS), and so on.

Despite the immense benefits of Nanosensors in agricultural and food processing, there is widespread public concern about toxicity and environmental impact. There is relatively little known about its long-term negative impact on soil, plants, and, eventually, humans.

Preliminary animal investigations have revealed that nanoparticles may be hazardous to the liver, kidneys, and immune system. Furthermore, the consequences of exposure to manufactured nanoparticles may differ from those generated by naturally occurring nanoparticles.

As a result, risk assessment studies to demonstrate the deleterious impacts of nanoparticles on human health and the environment should be standardized and expanded, while the subject of nanotoxicology arises at an international level. Clearly, knowledge of the existence of nanoparticles in the environment and their detection, measurement of nanoparticle emissions, life-cycle, toxicity, and influence on human health and the environment are all critical in realizing all of the benefits that nanotechnologies have to offer.

The toxicity of nanoparticles in the environment is determined by their size, kind, charge, and other factors. Furthermore, the impact of nanoparticles on the environment is affected by environmental conditions (humidity, temperature, wind flow rate, the nature of light, etc.).

However, the features of nanomaterials, such as their tiny size and huge surface area, allow for facile dispersion and bonding in the environment as well as with human tissues. Agriculture is one method by which nanomaterials reach the environment and humans.

Nanoparticles have significant interaction with soils. Nanomaterials augment plant meals, and pesticides can diffuse into the soil, water, and the atmosphere, forming stronger bonds with contaminants and transporting them via soil and water.

Exposure to nano fertilizers and pesticides can pose health risks. The migration of nanoparticles from food to humans is also a considerable concern. In other words, direct consumer exposure to nanoparticles presents a major threat to human health. Nanoparticles could enter the bloodstream through the skin, respiratory system, or digestive tract.

Therefore, exposure is restricted or extremely minimal as long as the nanoparticles stay bonded. Teow et al. (2011) published extensively on the health effects and safety of nanomaterials.

The utilization of biosensors in conjunction with enhanced technology in molecular biology, nanomaterials, and microfluidics has huge possibilities for agricultural yield. These are also used to track the activity of microorganisms in the soil and forecast the likelihood of soil diseases.

The main premise of soil inspection using a biosensor is to determine the comparative action of positive and negative bacteria in soil based on variations in oxygen utilization during their breathing. They also provide several alternatives for monitoring pollutants and their impediments by utilizing novel nanomaterial features.

Biosensors for detecting nitrate concentrations in plants, as well as markers to identify diseased plants, have been described for methyl salicylate and azelaic acid. It was stated that biosensors were used to monitor *Penicillium digitatum* infection in citrus fruits.

Smart delivery systems and Nanosensors are used to aid in the efficient use of natural agricultural resources such as nutrients, water, and chemicals through precision farming techniques such as satellite monitoring, geographic systems, and remote detecting tools that can detect pests on crops or indicators of stress such as drought.

The use of independent sensors linked to GPS tracking in real-time is believed to play an important role in nanotechnology-assisted instruments. Nanosensor placement may be done across the field to monitor crop development and soil factors.

Nanosensors are gaining popularity and playing an important part in the growth of the food sector due to their efficient response instruments for detecting gases, bacteria, or harmful compounds in packaged meals. Nanobiosensors have been tested for detecting infections in processing facilities as well as informing clients, procedures, and suppliers about food safety.

It has also been used to detect the presence of pollutants, mycotoxins, and microorganisms in food. Allergens have also been identified using biosensors and nanoparticles, and the report is set to be commercialized. These gadgets can also detect time, expiration date, and temperature history.

6.8. NEW OPPORTUNITIES IN AGRICULTURE

There have been at minimum two important components of primary production management where study can deliver significant breakthroughs to satisfy needs in the future:

- Better resource utilization efficiency,
- Improved resource utilization efficiency.

lack of specific statutory guidance. For example, in the European Union, an effort is being made to develop a legislative and regulatory structure.

6.9. THE CHALLENGES OF NANOTECHNOLOGY IN FOOD AND AGRICULTURE SYSTEM

As previously discussed, nanotechnology is widely employed for food and agricultural purposes. However, nanoparticles are related to several safety concerns due to their possible danger values infiltrating cells due to their small sizes and possibly existing in the system.

Because of the numerous antagonistic effects of various nanoparticles, the increased application of nanotechnology in agricultural operations and food items is of great concern among a large segment of society. Although the qualities and protection of substances in bulk are often obvious, nanoscale complements consistently exhibit different features than those at the macroscale.

The hazard posed by nanotechnology is mostly due to the tiny size of nanoparticles with greater surface surfaces that are highly dispersible, penetrating cells to reach distant locations of the body and potentially causing toxicity. Nanomaterials have the potential to interact with biological specimens due to their size resemblance to DNA.

Environmental conditions can cause nanocomposites to degrade, resulting in the release of incorporated nanoparticles from polymeric textures into the environment. Some food packaging, such as low-density polyethylene, loses strength after being exposed to environmental factors such as UV light or ozone in humid circumstances.

The large volumes of nanoparticles used in food packaging may be a source of concern due to their leakage from tainted food or the environment. However, there is a scarcity of evidence on the migration of nanoparticles from packaging materials into food, as well as their toxicological effects.

The use of green nanofillers in nanocomposites research is critical for the protection of the environment, human health, and animals. Furthermore, temperature, pH value, polymeric structure and viscosity, contact duration, mechanical pressure, and food content are important considerations.

The exact estimation of nanoparticles emitted into the environment, their accumulation in plants, and their influence on health are equally critical to analyze.

6.10. CONCLUSION

The public concern regarding health risks associated with nanotechnology products is questioned by many scientific communities and needs further investigation. It is a developing and quickly growing region with new and exclusive applications in agriculture and food research.

Compared to other fields of nanotechnology application, like medicine, materials, and energy, agriculture is still a marginal sector. Food nanotechnology comprises the application of nanocarrier methodologies to strengthen the bioactive ingredients to modify their biological accessibility and barrier against several chemical or environmental variations.

Many research investigations are focusing on increasing the application of nanotechnology for the production of crops and food processing. Nanotechnology-based sensors for trace detection, observing the condition of diet in terms of transportation and storing, and encapsulation of food modifiers or additive materials were other merits of the technology.

Agricultural efficiency, soil improvement, secure water use, distribution of food in stores, and its quality are basic factors of securing food that may be improved via advances in nanotechnology research. In more general terms, the application of nanotechnology in plant production systems or—more broadly—in plant science was defined with the term "Phyto nanotechnology."

Nanotechnology is benefiting food industries as novel food packing supplies with modified mechanical, fencing, and antimicrobial characteristics.

Newer technology that will increase production and reduce food wastage is important to maintain the sustainable living standards of the nation and improve food security. Nanotechnology can provide a path for producing foods with outstanding quality in highly improved workable form along with induction of nutrients bioavailability.

Farmers throughout the world will focus on using new innovations and technologies to enhance crop production through intensive and extensive agriculture.

REFERENCES

1. Banigo, A., Azeez, T., Ejeta, K., Lateef, A. and Ajuogu, E., 2020. Nanobiosensors: applications in biomedical technology. *IOP Conference Series: Materials Science and Engineering*, [online] 805(1), p.012028. Available at: <https://www.researchgate.net/publication/341832901_Nanobiosensors_applications_in_biomedical_technology> [Accessed 2 June 2022].
2. Dr. Priyom Bose, P., 2020. *The Role of Nanotechnology in Ensuring Food Safety*. [online] AZoNano.com. Available at: <https://www.azonano.com/article.aspx?ArticleID=5496#:~:text=In%20food%20processing%20industries%2C%20nanoparticles,fresh%20for%20a%20longer%20duration> [Accessed 2 June 2022].
3. Marchiol, L., 2018. Nanotechnology in Agriculture: New Opportunities and Perspectives. *New Visions in Plant Science*, [online] Available at: <https://www.intechopen.com/chapters/61312> [Accessed 2 June 2022].
4. Neme, K., Nafady, A., Uddin, S. and Tola, Y., 2021. Application of nanotechnology in agriculture, postharvest loss reduction and food processing: food security implication and challenges. *Heliyon*, [online] 7(12), p.e08539. Available at: <https://www.ncbi.nlm.nih.gov/pmc/articles/PMC8661015/> [Accessed 2 June 2022].
5. Neme, K., Nafady, A., Uddin, S. and Tola, Y., 2021. Application of nanotechnology in agriculture, postharvest loss reduction and food processing: food security implication and challenges. *Heliyon*, [online] 7(12), p.e08539. Available at: <https://www.sciencedirect.com/science/article/pii/S2405844021026426> [Accessed 2 June 2022].
6. Omanović-Mikličanina, E. and Maksimović, M., 2016. *Nanosensors applications in agriculture and food industry*. [ebook] http://hemija.pmf.unsa.ba, pp.59 - 70. Available at: <http://hemija.pmf.unsa.ba/glasnik/files/Issue%2047/5-59-70-Omanovic.pdf> [Accessed 2 June 2022].
7. Sharma, P., Pandey, V., Sharma, M. and Singh, B., 2021. A Review on Biosensors and Nanosensors Application in Agroecosystems. *Nanoscale Research Letters*, [online] 16(1). Available at: <https://nanoscalereslett.springeropen.com/articles/10.1186/s11671-021-03593-0#article-info> [Accessed 2 June 2022].

CHAPTER 7

MICROALGAL BIOTECHNOLOGY: PROSPECTS AND APPLICATIONS

CONTENTS

There is a strong international focus on discovering novel and safer antioxidants using plant sources such as plant material to avoid oxidative degradation in food and also to reduce oxidative harm to living cells. Microalgae are photosynthetic microorganisms that could rapidly build cellulose from solar energy, CO2, and nutrients in water bodies.

Such biomass contains essential primary metabolites such as sugars, oils, and lipids, for which process pathways exist for the creation of high-value products such as human and animal feed supplements, transportation fuels, industrial chemicals, and medicines.

7.1. INTRODUCTION

Algal biomass and algae-derived chemicals have a wide range of potential uses, ranging from animal feed and aquaculture to human nutrition and health products. A few algae are thought to be high in natural antioxidants. Despite the fact that macroalgae have gotten a lot of interest as possible natural antioxidants.

Furthermore, by employing clean nutrient media for growing the microalgae, the characteristics of the microalgal cells may be managed such that they contain no herbicides, pesticides, or other hazardous compounds. The simplicity with which specific components may be purified adds to the potential of microalgae as a source of natural antioxidants.

There have been few studies on the antioxidant properties of microalgae. Cyanobacteria provide a tremendous possibility for discovery since they are essentially unknown; the projected rate of rediscovery is considerably lower than for other better-studied groups of organisms.

Beneficial chemicals are the byproducts of secondary metabolism, which occurs when conditions are not favorable for rapid development. To manufacture such compounds, new strains (faster growth, greater substrate tolerance, etc.) must be developed by classical selection or genetic modification so that microalgal biomass can be produced regularly.

To generate healthy organic food, emphasize the importance of dietary antioxidants and their potential advantages in health and disease, either directly or indirectly through plant nutrients and animal feed.

Study the many biological activities of algae, including their relationships with its biochemical makeup, pigments, and components, which may alter with salt-stressed growing circumstances, and define the antioxidant properties of algae.

7.1.1. Microalgae

Microalgae are photosynthetic prokaryotic or eukaryotic microorganisms that create carbohydrates, proteins, and lipids through photosynthesis. Because of their unicellular or basic multicellular structure, they may develop quickly and survive in adverse environments.

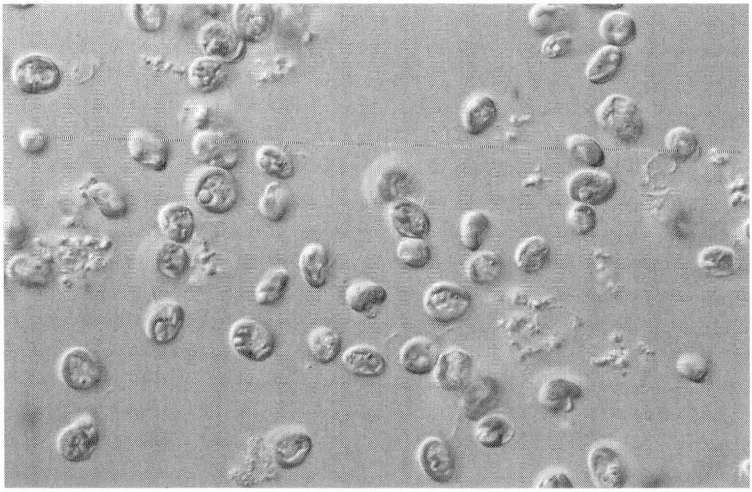

Figure 7.1. An Image of Microalgae.

Source: Image by Wikimedia Commons

Cyanobacteria (*Cyanophyceae*) are examples of prokaryotic microorganisms, whereas green algae (*Chlorophyta*) and diatoms are examples of eukaryotic microalgae (*Bacillariophyta*). Microalgae are found in all current earth ecosystems, both aquatic and terrestrial, and represent a diverse spectrum of organisms that live in a variety of environmental situations.

More than 50,000 species are thought to exist, although only a small fraction, approximately 30,000, have been researched and investigated. The main prerequisites for developing algae are sunlight, water, nutrients, and fertile land. Microalgae have the potential to repair CO2 using solar energy with ten times the efficiency of terrestrial plants, as well as various other technical benefits.

Algae use sunlight more efficiently than terrestrial plants, absorb toxic contaminants, have low resource requirements, and do not compete for scarce resources with food or agriculture.

7.1.2. Algal Metabolites

Metabolites are metabolic intermediates and products. Small molecules are typically referred to as metabolites. A primary metabolite is a substance that is directly engaged in proper growth, development, and reproduction.

Secondary metabolites are not directly engaged in these activities, but they generally serve a vital ecological purpose. Secondary metabolism is induced in response to certain environmental variables or developmental phases. Secondary metabolites are chemical molecules found in organisms that are not directly involved in the organism's regular growth, development, or reproduction.

Research of these species for pharmaceutical purposes has provided crucial chemical prototypes for the identification of novel drugs, encouraging the employment of advanced physical methods and innovative syntheses of medicinal chemicals. Secondary and primary metabolisms have been explored in this regard as a forerunner to future sensible economic exploitation.

The secondary metabolism has a narrow distribution, whereas the main metabolism provides intermediates for the production of important macromolecules.

7.1.3. Phytochemicals

The Greek term for a plant is "Phyto." The word "phytochemicals" refers to a wide range of plant-produced substances. Phytochemicals are chemical substances produced by plants during normal metabolic processes. There are various "families" of phytochemicals, and they all benefit the human body in different ways.

Phytochemicals have the potential to protect humans against a variety of ailments. Such molecules are known as "secondary metabolites," and there are various kinds of them, including alkaloids, flavonoids, coumarins, glycosides, gums, polysaccharides, phenols, tannins, terpenes, and terpenoids.

Phytochemicals are non-nutritive substances that exist naturally. They appear to act both alone and in conjunction, maybe in combination with vitamins.

Bioactive chemicals have been present in plants from their inception. However, their truth has been recognized for hundreds of years. Herbal plants are commonly utilized all over the world since they are the primary source of phytochemicals. It is thought that understanding of traditional remedies arose via the trial and error concept over generations.

According to extant records, the Chinese have the world's oldest medicine and treatment system. Over five thousand years ago, the Chinese used their medications because of the effects of "yang" and "yin," as well as the 05 components. Herbal remedies have been used since 2800 BC, when the Chinese ruler "Shen Nong" penned the book. "The Great Indigenous Herbal." Similarly, the renowned scientists Aristotle and Hippocrates were the first to develop herbal medications for the treatment of deadly diseases. During the 1800s and 1900s, the primary focus of researchers was to discover active molecules with pesticide or therapeutic qualities, such as morphine, salicylic acid, and pyrethroids (pesticides).

Many laboratories were built in the 1980s to discover phytochemicals in plants that may be utilized as medications. Many of these newly found phytochemicals are used to treat diseases and infections such as cancer, heart disease, and stroke. Simultaneously, other researchers did epidemiological research to determine the association between the use of certain phytoconstituents and human health.

Recent research has shown that a daily diet high in dietary plants might reduce the risk of chronic diseases such as cardiovascular disease, type 2 diabetes, neurodegeneration, tumor development, and cancer of various sorts to a larger extent.

Phytochemicals are found in appropriate amounts in nuts, fruits, whole grains, beans, seeds, and vegetables. It has been stated that consuming fruits high in phytochemicals may help prevent at least one in every five (20%) incidences of cancer, as well as other major ailments such as heart disease. Phytochemicals aid in the prevention and treatment of the following diseases and ailments. Phytochemicals boost the body's immune system against disease-causing organisms such as bacteria, fungus, viruses, and so on. Inhibit the formation of cancer-causing compounds (carcinogens) in the body as a result of substances we consume from the environment in the form of food or drinks.

Reduce oxidation, the cell damage caused by aging, and environmental exposure. Free radicals are responsible for oxidation, which causes cell abnormalities and, as a result, may lead to cancer.

- Reduce the pace at which cancerous cells develop.
- Reduce the soreness that contributes to cancer development.

- Activate the process of cell death (apoptosis) in wounded cells that may be precursors to cancer.
- Not only does it prevent DNA damage, but it also aids in its repair mechanism.
- Allowing for the control of hormones such as insulin and Oestrogen in the body. An overabundance of these hormones raises the risk of colon and breast cancer.

7.2. MICROALGAL BIOACTIVE COMPOUNDS

Microalgae are indeed a rich source of bioactive metabolites, notably cytotoxic chemicals used in cancer treatment. Antibiotic compounds were identified in marine microalgae, such as *Phaeocystis sp. blooms*. Acrylic acid, which accounts for around 7.0 percent of the dry weight, is claimed to be produced by *Phaeocystispouchitis*.

Antibiotic compounds are created in this manner and are discovered in the digestive tracts of Antarctic penguins. The halotolerant alga *Dunaliella sp.* has been shown to produce ß carotene and vitamins. These chemicals are extremely important in mariculture operations.

Cyanobacteria have been recognized as one of the most promising groups of organisms for the isolation of innovative and biochemically active natural compounds. Secondary metabolites are produced by cyanobacteria such as *Spirulina, Anabaena, Nostoc*, and *Oscillatoria.*

Cyanobacteria create a wide range of bioactive substances, including 40% lipopeptides, 5.6 percent amino acids, 4.2% fatty acids, 4.2% macrolides, and 9% amides.

Figure 7.2. Bloom of cyanobacteria in a freshwater pond.

Source: Image by Wikimedia Commons

Cyanobacterial lipopeptides contain various compounds such as cytotoxic (41%), antitumor (13%), antiviral (4%), antibiotics (12%), and the remaining 18% activities include antimalarial, antimycotics, multi-drug resistance reversers, antifeedant, herbicides, and immunosuppressive agents; in addition to the immune effect, blue-green algae improve metabolism.

Cyanobacteria are also known to create anticancer, antiviral, and antifungal chemicals, as well as to decrease cholesterol in both humans and animals. Most of those pharmaceutically relevant chemicals found in cyanobacteria are peptides, including cyanobacterial toxins and anticancer therapeutic possibilities.

Peptide synthetases are widely found in cyanobacteria and are responsible for the synthesis of cyanobacterial hepatotoxins and other peptides. Certain cyanobacterial bioactive chemicals are also biosynthesized by polyketide synthetases (e.g., microcystins). Several extracts were discovered to be quite effective in shielding human lymphoblastoid T-cells against the cytopathic consequences of HIV infection.

Active compounds derived from *Lyngbyalagerheimii* and *Phormidium tenue* included sulfolipids with various fatty acid esters. Cyanovirin is a protein obtained from an aqueous cellular extract of *Nostoc elipsosporum* that inhibits primate retrovirus multiplication and pathology *in vitro*. Cryptophycinseems to be an active chemical identified from the Nostoc strain that exhibits antiproliferative and antimitotic properties by attaching to the ends of microtubules and therefore interrupting the cell cycle at the mitotic metaphase.

A synthetic counterpart, cryptophycin-52, is now in Phase II clinical studies due to its significant anticancer efficacy. Calcium spirulan, sulfated polysaccharide A new antiviral agent has been discovered in water-soluble extracts of cyanobacteria. This chemical appears to preferentially impede encapsulated viral entry into host cells, hence halting reproduction.

The impact has been documented on a variety of viruses, including herpes simplex, measles, and even HIV1. Among eukaryotic microalgae, a glycoprotein derived from *Chlorella vulgaris* culture supernatant protected mice from tumor spread and chemotherapy-induced immunosuppression.

7.2.1. Fatty Acids (Fa)

Microalgae have a specific lipid composition, particularly fatty acids such as EPA and DHA, which have economic importance. Microalgae's capacity to synthesize and accumulate a high number of PUFAs makes it even more valuable as a nutraceutical.

Omega-3 fatty acids are widely recognized and recommended as an excellent source of dietary supplements for their health advantages, particularly in disease prevention and human nutrition. Algal oil is commonly utilized in liquid or pill form and aids vegetarians as well as populations with poor marine diets.

The use of algal oils in food became widespread when advances in microencapsulation and refining technology eliminated any off-flavors in the food when coupled with the oils. Algal oils are supplemented or fortified in a variety of meals, including dairy products, nutritional bars, bread items, and so on, to boost the nutritional content of the product.

Microalgae generate total lipid up to 30–70% of dry mass based on the species, which serves as an energy reserve under harsh conditions and cell division. Fatty acids are components of complex lipids, and their structural diversity contributes to their numerous advantages.

Fatty acids are classified into two types based on the polarity of their molecular heads: (i) neutral lipids and (ii) polar lipids. Fatty acid synthesis in microalgae may be boosted by adjusting several environmental parameters such as oxygen level, temperature, light exposure, pH, and nutrition supplementation.

A number of studies have demonstrated that nitrogen deficiency increases the production of neutral lipids and triacylglycerides (TAG) in microalgae. Similarly, exposure to low light intensity and hence low temperature has been shown to increase PUFA synthesis.

Chlorella sp. cultivation under low CO2 conditions promoted high alpha-linolenate fatty acid content, but *Chlamydomonas reinhardtii* mutant CIA-3 displayed increased PUFA content in high CO2 conditions.

Mechanical pressing, homogenization, milling, and solvent extraction are all methods for extracting lipids from microalgae. The solvent extraction is critical for determining the polarity or solubility of the lipid of interest. To aid lipid extraction by solvent, other procedures such as enzymatic extraction, supercritical extraction, ultrasonic-assisted extraction, and microwaves are utilized.

Nonetheless, just a few cell disruption procedures, particularly steam explosion, enzymatic hydrolysis, bead milling, and horn sonication, are commercially viable. Purification of algal oil (PUFAs) is an important step since the crude form of these oils is inedible due to contaminants, odor, and taste, and it is unappealing for ingestion due to its muddy appearance.

Because oil is vulnerable to oxidation, the speed of operation and process condition are essential elements in oil purification processes. Martek Bioscience Corporation, for example, has documented the oil recovery and purification procedures of DHA from algal oil.

For example, a protease enzyme was employed to break the protein in *Schizochytrium sp.* cells, releasing the oil into the culture broth and generating an emulsion; isopropyl alcohol was then added to separate the oil. The hexane solvent extraction technique was used for *Crypthecodinium cohnii* since the enzyme approach is incapable of hydrolyzing the algal cellulose layer.

Centrifugation was used to remove the cell walls following solvent extraction, and the oil was recovered after evaporation of the solvent.

7.2.2. Sterols

Sterols are found in plant, animal, and microbe membranes and are known as phytosterols, zoosterols, and mycosterols, respectively. Cholesterol is the most prevalent zoosterol, but sterols in plants are usually found in mixes, with -sitosterol, campesterol, and stigmasterol being three of the most frequent phytosterols.

These sterols are all 5-sterols; however, 7-sterols may be found in trace amounts. Sterols are molecules with a sterol-like structure but no endocyclic double bond. Methyl sterols, also known as triterpenyl alcohol, feature an extra methyl group at carbon-4 in the molecule's A-ring. In-plant membranes, methyl sterols, and dimethylsterols coexist alongside sterols.

Bacterial membranes typically lack sterols, whereas yeasts collect significant amounts of sterols, which can account for up to 10% of the cellular dry weight. Algae generate a diverse range of sterols.

Sterols are a significant chemical element of microalgae. Sterols are the primary constituents of eukaryote life, and the sterol patterns of various groups of species differ. Sterols, as a result, serve as a fingerprint for organic materials introduced into an aquatic environment.

Furthermore, when deposited in anoxic sediments, sterols have a relatively high resistance to degradation and remain in the environment for a longer duration of time. Because it comes from a number of sources, cholesterol is the most abundant and pervasive sterol molecule in the environment.

The majority of physiologically generated sterols are planar 3-hydroxy tetracyclic structures with a methyl- or ethyl-substituted C7-C11 hydrocarbon side chain and a variety of methyl-substitution (C4, C14) patterns on the polycyclic nucleus with various degrees and places of unsaturation (C5, C7, C8).

The sterols' rigid structure, created by the fused ring system, offers cell membrane integrity and stability, holding the membrane together. In general, there is no one sterol that can be attributed to a single algal source. Many of the sterols listed above can also be found in other types of algae.

7.2.3. Pigmentation in Aquacultures

Astaxanthin is a red pigment found in a variety of aquatic creatures, including microalgae, seagrasses, shrimp, lobsters, salmon, and trout. Crustaceans are unable to synthesize carotenoids from scratch and must consume astaxanthin (or sufficient precursors) to get the required hue for seafood market acceptability.

For this aim, natural sources such as the algae *Dunaliella salina* and *Spirulina maxima*, as well as synthetic ones like β-carotene, canthaxanthin, and astaxanthin, have been employed. In fact, astaxanthin is one of the most expensive components of salmon aquaculture, accounting for around 15% of overall production cxpcnses.

Biotechnology firms have investigated the green unicellular freshwater alga *HaematococcusPluvialis* as one of the numerous natural sources of astaxanthin used in aquaculture. β-carotene is an important component of the carotenoid family, which is a group of natural fat-soluble stereoisometric pigments.

β -carotene has pro-vitamin A activity and hence serves a vital function in the human body. β -carotene could also be used to color foods. As a result, β-carotene has a wide range of uses in food, medicines, and cosmetics. The industry has fulfilled the high demand for -carotene, primarily through synthetic synthesis. The increased demand for natural carotenoids has fuelled interest in extracting -carotene from various natural sources. *Dunaliella salina* is the market's primary source of natural β -carotene.

Natural β -carotene has a market size of 10-100 tonnes. Per year-1 and a price of >750 €. Kg-1. Furthermore, β -carotene (like other carotenoids) is a powerful antioxidant, scavenging potentially harmful oxy radicals that have been linked to the induction of certain cancers, and there is an inverse relationship between the consumption of certain carotenoids and the risk of cancer.

Carotenoids' antioxidant activity is the foundation of their anti-oxidative stress protection in various species and circumstances. Carotenoids' effects on human health are often linked to their antioxidant qualities. Nonetheless, not all of the biological activities attributed to carotenoids must be connected to their capacity to limit the buildup of free radicals and reactive oxygen species.

Dunaliella salina, a halophilic green biflagellate microalga, has long been known as an effective biological source of this carotenoid. Many epidemiological studies indicate that individuals given a diet high in β-carotene from *Dunaliella*, which maintains higher than normal blood carotenoid levels, have a decreased incidence of numerous forms of cancer and degenerative disorders.

Here are several uses for xanthophyll astaxanthin in nutraceuticals, cosmetics, and the food and feed sectors. A number of other possible uses for this carotenoid, primarily linked to human health and nutrition aspects, have lately been suggested.

Lutein is one of the most abundant carotenoids in foods and human blood, and it is a necessary component of the pigment found in the macula lutea (or yellow spot) of the retina and the lens of the eye. Lutein is utilized as a food color as well as a feed ingredient in aquaculture and poultry husbandry.

Additional uses for lutein have sparked substantial attention in recent years, particularly those connected to human health.Lutein is currently thought to be an effective agent for the preventative measures of a range of human diseases, primarily based on epidemiological studies.

The microalga *Muriellopsis sp.*, as well as other chlorophycean species, may store lutein in their biomass. There is no known commercial method for producing lutein from microalgae; however, the foundation for outdoor production of lutein-rich cells of *Muriellopsis* and *Scenedesmus* strains on a trial scale has previously been synthesized.

7.4.3. Antimicrobial Activity

Microalgae antibacterial activity has been linked to molecules from a variety of chemical families, including indoles, terpenes, acetogenins, phenols, fatty acids, and volatile halogenated hydrocarbons. The antibacterial activity of supercritical extracts produced from the microalga *Chaetoceros muelleri*, for example, was linked to its lipid content.

The antibacterial action identified in different compressed extracts of *Dunaliella salina*, however, may be explained not only by fatty acids but also by substances such as - and -ionone, -cyclocitral, neophytadiene, and phytol.

Research is needed to identify the molecules directly responsible for such antimicrobial properties, such as chlorellin, which has been underway but is still in its early stages due to the discovery of several new families of compounds.

Microalgal cell-free extracts are already being studied as food and feed additives in an effort to replace synthetic antimicrobial substances already in use – including subtherapeutic dosages of antibiotics used as a prophylactic strategy in animal breeding.

7.4.3. Antiviral Activity

Virus-borne contagious diseases have developed (and re-appeared) in recent years. Despite the fact that various antiviral medications have been specifically produced, drug-resistant mutations are continually occurring, necessitating the development of novel antiviral active principles, particularly those derived from sources that do not form (or are exposed to) viral pools.

This is why microalgae have attracted considerable interest as prospective providers of antiviral medicines. Viral growth is often separated into three phases, with antiviral activity occurring at one or more of these stages:

Stage I comprises cell adsorption and invasion

Stage II, or the eclipse phase, in which the cell is pushed to synthesize numerous copies of the virus; and

Stage III, or viral particle maturation and release.

For example, the antiviral drug acyclovir® expresses anti-HSV activity at stage II during the anti-HSV factor from *Dunaliella* sp.

At stage I, it deactivates the viral function. Sulfated exopolysaccharides from marine microalgae have been reported to interfere with Stage I of various enveloped viruses; they provide competitive advantages due to their broad antiviral range against, for example, influenza.

HIV-1 and HSV Their inhibitory impact appears to result from contact with positive charges on the virus or on the cell surface, stops the former from penetrating host cells; they may also selectively block reverse transcriptase in the case of HIV, so delaying the creation of new viral particles following infection; however, the precise stage of viral replication when they act remains unknown.

Antiviral highly sulfated polysaccharides derived from many species of red microalgae are mostly composed of xylose, glucose, and galactose; they are extremely stable when subjected to severe pH and temperature.

Considering their antiviral effectiveness, the metabolic mechanisms that lead to sulfated polysaccharides are yet unknown. The secretion of unicellular red algae was first studied using radiolabeling, which revealed that biosynthesis of the carbon chain and sulfation of the resultant polysaccharide occurs in the Golgi apparatus; these findings were verified in *Porphyridium sp.* and other red microalgae.

Quite lately, researchers have used 14C pulse-chase experiments and ultra-structural microscopy to reach the conclusion that brefeldin A–a Golgi apparatus membrane-traffic inhibitor–decreases the contents of the bound and soluble forms of polysaccharides while inhibiting cell-wall binding of polysaccharides more than its soluble counterpart (in both actively growing and resting cells).

The discovery of tiny compounds that can selectively disrupt a specific protein-protein interface remains a problem, but it is of particular importance in virology because the present antiviral medications primarily target viral proteins.

7.4.4. Antibacterial Activity

Secondary metabolites are produced by microalgae and accumulate in the growth media towards the conclusion of the exponential and stationary phases of development. Antibacterial action has been suggested for a range of substances, including fatty acids.

When compared to the control treatments, all microalgae cultures suppressed bacterial growth ($P < 0.05$), and their antibacterial activity was unaffected by the presence or absence of light. In the absence of microalgae cells, the number of bacteria in the control groups expanded rapidly over the trial period, suggesting that the bacterial cells were able to use the growth medium of microalgae cultures.

The current findings may explain the low levels or lack of Vibrio strains in microalgae cells, as well as the beneficial effect of the addition of microalgae in fish larval rearing and the synthesis of antimicrobial chemicals by microalgae cells.

The majority of efforts were dedicated to the research of antibiotic-resistant bacteria for several purposes:

(i) Bacterial infections are responsible for the majority of community-acquired and nosocomial infections,
(ii) The large and expanding number of antibacterial classes offers a more diverse range of resistance mechanisms, and
(iii) The large and growing percentage of antibacterial courses provides a much wider variety of resistance mechanisms, and
(iv) The capacity to transfer bacterial resistance determinants into standard, well-characterized bacterial strains allow for more in-depth research of the underlying molecular processes.

Pratt et al. discovered the first antibacterial chemical from Chlorella, a microalga; a combination of fatty acids, Chlorellin, in particular, was seen to be essential for such an inhibitory effect against both Gram+ and Grambacteria.

However, the production of active antibacterial principles by microalgae has skyrocketed. This realization occurred, for example, from the risk associated with certain multidrug-resistant *Staphylococcus aureus* (MRSA) strains, which have caused significant worry in healthcare facilities globally – since they are resistant to most conventional medications. As a result, the development of new antibacterial agents with diverse biochemical modes of action is encouraged.

Antibiotics are often less effective against Gram-negative bacteria because of their complex, multilayered cell wall construction, making it much more difficult for the active component to enter. This explains why the supernatant (and methanolic extracts) are more effective against Gram+ bacteria than Gram- bacteria.

The precise mechanism of action of fatty acids is unknown: they may act on multiple cellular targets, though cell membranes are the most likely – as membrane damage will likely result in cell leakage and reduced nutrient uptake, in addition to inhibiting cellular respiration; on the other hand, Desbois claimed a peroxidative process.

Moreover, chemicals generated by *Scenedesmus costatum* and partly isolated from its organic extract displayed efficacy against aquaculture bacteria due to fatty acids with chains longer than 10 carbon atoms – which appear to promote bacterial protoplast lysis.

The capacity of fatty acids in general to inhibit bacterial growth and survival has long been recognized, but new structure-function connection studies reveal that this ability is dependent on both chain length and degree of unsaturation. Since substances like cholesterol can counteract antibacterial properties, both the content and concentration of free lipids must be considered.

The antibacterial properties of polyunsaturated aldehydes merit special mention among microalgal-derived oxylipins. Diatoms, such as *S. costatum* and *Thalassiosira rotula*, produce such chemicals.

7.4.5. Antifungal Activity

Algae are among the most important biological agents explored for such control of fungus and plant diseases. Cyanobacteria strains are expected to cause intracellular and extracellular metabolites with a wide range of biological activities, including antibacterial, antifungal, and antiviral activity. Antibiotics and poisons are examples of physiologically active chemicals.

Hussien et al. investigated the effect of culture filtrates from nine algal strains (*Anabaena flosaquae, Anabaena oryzae, Chlorella vulgaris, Nostoc muscorum, Nostoc humifusu, Oscillatoria sp., Phormedium fragile, Spirulina platensis,* and *Wolleasaccata*) at concentrations of 10, 20, 30, and 40% (100, 200, 300 and 400 ppm, respectively).

Researchers discovered that, generally, all algal culture filtrates inhibited fungal mycelium development, although *Spirulina platensis, Oscillatoria sp.*, and *Nostoc muscorum* attained the maximum fungal mycelium growth inhibition percentages at concentrations of 30% (and 40%.

Topsin M70, a synthetic chemical fungicide, totally prevented fungal mycelium development at doses of 200, 300, and 400 ppm. The preceding three algal culture filtrates at 300 and 400 ppm, notably at 40% concentration, fully prevented fungal spore formation (number of spores). Topsin M70 had the same results.

The existence of bioactive substances in the algal culture filtrates, such as total phenolic compounds, total saponins, and alkaloids, has been related to the antifungal activity of the algal culture filtrates.

7.4.6. Nematicidal Activity

Meloidogyne spp., root-knot nematodes, are among the most devastating worms in crops, generating an estimated $100 billion in losses each year. Chemical nematicides have been discovered to be an efficient method of controlling nematodes, but they have a hazardous aftereffect on the environment, notably on non-target creatures and human health.

Furthermore, the use of chemical nematicides is forbidden in organic farming. Except for a few observations showing endospores of *Microcoleus* and *Oscillatoria spp*. killed nematodes, the nematocidal potential of cyanobacteria has remained unknown.

Microcoleusvaginatus culture filtrates prevented the hatching of *Meloidogyne incognita* eggs and killed second-stage juveniles. Microalgal metabolites have piqued the interest of researchers since they represent a source of poisons and possible novel medications.

Figure 4.4. *Meloidogyne spp. at Solanum lycopersicum.*

Source: Image by Wikimedia Commons

The nematicidal effect of nine algal culture filtrates (*Nostoc muscorum, Anabaena flos aquae, Anabaena oryzae, Chlorella vulgaris, Wolleasaccata, Phormedium fragile, Oscillatoria sp., Nostoc humifusum,* and *Spirulina platensis*), *Azolla pinnata* aqueous extract filtrate (1:2 The laboratory experiment demonstrated that a substantial proportion of juvenile death occurred overall exposure times and treatments, with the greatest outcomes obtained after 72 hours of exposure.

At the highest concentration of 1:10, only five cyanobacterial strains, namely *Spirulina platensis, Oscillatoria sp., Anabaena oryzae, Nostoc muscorum,* and *Phormedium fragile*, significantly enhanced juvenile mortality by more than 70%. (84.3, 80.4, 78.9, 75.4, 72.5 and 70.1 per cent, respectively).

7.4.7. Mollscicidal Activity

Because the intermediate snail hosts of schistosomiasis are the locations of high parasite growth, their management efforts are prioritized in the decrease schistosomiasis transmission. However, chemical molluscicides are effective in controlling the disease in some cases.

However, due to their negative consequences, interest in ecologically friendly techniques and the employment of biological control agents has resurfaced.

The algal culture filtrate of *Spirulina platensis* was demonstrated to have molluscicidal activity against *Biomphalariaalexandrina* snails, an intermediate host of *Schistosoma mansoni* in Egypt, which supports one's survival and fertility, as well as an impact on the hatching abilityof snail eggs and also the survivability of the parasite's free-living larval stages (miracidia and cercariae).

With 90 LC 0.23 percent of the filter, the cultured filtrate was deadly to snails. After one week of continuous exposure to sublethal doses of 0.19 and 0.17 percent, *B. alexandrina* snails ceased producing eggs, whereas those subjected to 0.07 percent laid few.

It might be because of the phytochemical contents of the culture filtrate, which include total phenolic compounds, alkaloids, and total saponins. Histological study revealed substantial damage in the hermaphrodite gland cells of snails exposed to these concentrations.

B. alexandrina eggs aged 3 and 7 days failed to hatch following exposure to 0.1 percent algal filtrate, while 22 and 10% only hatched after exposure to 0.05 percent of the same filtrate, respectively. Freestanding cell culture filtration has significant miracidicidal and cercaricidal activities, with 2 percent of this filtrate killing the majority of these organisms after 15 minutes of exposure.

7.5. BIOTECHNOLOGICAL APPLICATIONS OF MICROALGAE

Algae are lower plants that lack features like leaves, roots, and stalks. Algae are not included in the Kingdom Plantae in recent categorization schemes based on molecular phylogenetic systematics, despite their close resemblance to higher plants.

Algae are classified as microscopic (microalgae) or macroscopic (macroalgae), with the latter consisting primarily of seaweeds.

Algae vary in size from picoplankton with diameters of about 0.2 – 2.0 mm to huge kelps with fronds up to 60 m in length. The number of algal species is believed to be one to 10 million, the majority of which are microalgae. Cyanobacteria are distinguishable from other bacteria by their ability to perform oxygenic photosynthesis.

Cyanobacteria were the first living forms to invade Earth, and through photosynthesis, they created oxygen, allowing other life forms to thrive. Based on their color, the algae are further classified into three broad groups: green, brown, and red algae.

7.5.1. Applications of Microalgae in Environmental Biotechnology

Microalgae have a wide range of uses in environmental biotechnology, particularly in bioremediation, bioassay, and biomonitoring of environmental toxins. The discharge of wastewater carrying different toxins into aquatic environments is a major source of worry since it can endanger public health.

The high-rate algae pond (HRAP) system has been demonstrated to be an efficient wastewater treatment technology. The system comprises a shallow pond mixed with paddlewheels to improve nutrient transfer and photosynthetic efficiency in order to maximize the growth of algae.

The HRAP method is well suited to tropical climates, which have warm temperatures and plenty of sunlight all year. Aside from efficiently reducing pollutants such as chemical oxygen demand (COD), biological oxygen demand (BOD), nitrogen, and phosphorus, HRAP creates algal biomass, which has potential applications as animal feed and biodiesel feedstock.

Microalgae cultivated in HRAP have been proven to be effective in the treatment of a variety of wastewaters, including rubber effluent, palm oil mill effluent (POME), and municipal wastewater. *Spirulina platensis* cultivated in HRAP reduced phosphate from anaerobically digested starch manufacturing effluent by more than 99 percent.

The HRAP technique is helpful for tertiary wastewater treatment following conventional treatment before discharge. *Chlorella vulgaris* cultivated in HRAP, for example, has been proven to be beneficial for the final cleaning of textile effluent before release, particularly for color removal.

In addition to single-species cultivation, a mixture of multiple microalgae has been examined in several studies for wastewater treatment efficiency. For example, a consortium of five HRAP-grown microalgae was effectively employed to remediate landfill leachate.

In addition to suspended cultures, an immobilized microalgae system can improve the effectiveness of toxicant elimination in the environment. *Chlorella vulgaris* immobilized in alginate, for example, was shown to be useful in removing color from textile colors.

Immobilized *Chlorella vulgaris* and *Scenedesmus obliquus* have been found to be efficient in removing nitrogen and phosphorus from semicontinuous urban effluent. Co-immobilization of microalgae and bacteria has been demonstrated to be more successful than immobilized microalgae alone in removing nutrients such as nitrate, ammonium, and phosphate from wastewater.

Although toxin-tolerant microalgae are valuable in bioremediation, sensitive species are excellent for bioassay and biomonitoring of environmental contaminants. To investigate the toxicity of contaminants such as heavy metals, herbicides, and medications, microalgae have been utilized as bioassay organisms.

Heavy metals have been found in aquatic habitats as well as marine species like fish and mussels. Because of intensive agricultural and urban activities, residual organochlorine and organophosphate pesticides have been identified in the Selangor River.

7.6. OTHER APPLICATIONS AND PRODUCTS FROM MICROALGAE

Commercial applications for microalgae include natural supplies of important macromolecules such as carotenoids, long-chain polyunsaturated fatty acids, and phycocolloids. Because photoautotrophs have minimal growth requirements, they are appealing to bioprocesses aiming at creating high added-value chemicals in great demand by the pharmaceutical industry.

Microalgae production and biochemical composition are substantially influenced by growth method, medium composition, and nutrient profile. As a result, several initiatives have been established to elucidate the practical implications of the aforementioned factors. As a result, there is a rising interest in the study of the favorable effects of algae on human health and other advantages.

7.7. USE OF MICROALGAE FOR BIODIESEL PRODUCTION

In recent years, there has been considerable interest in the possible use of microalgae as a feedstock for biodiesel synthesis. When compared to other agricultural plants, using microalgae for biodiesel production has the benefit of not interfering with the production of food, fodder, and other feedstocks generated from such crops.

Palm oil has always been the main raw material for biodiesel manufacturing in Malaysia; nevertheless, availability is insufficient to fulfill demand. Thereis concern that if palm oil is consumed for biodiesel production, there might not be much left for food cultivation.

Microalgae seem to become the only biodiesel source with the ability to replace fossil fuels. Microalgae oil concentration can range from 16 to 68 percent dry weight. In comparison to other plant crops, which vary from 172 to 5950 L/ha, the oil output from microalgae may reach up to 136,900 L/ha. Manipulation of cultivation conditions can increase the lipid content of microalgae even further.

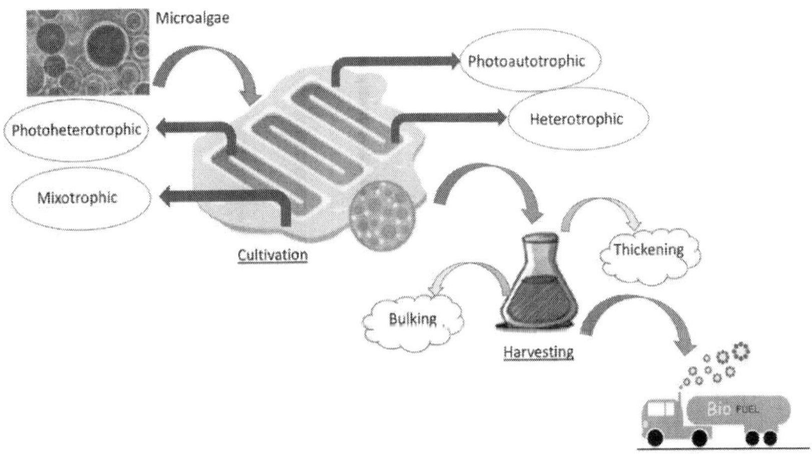

Figure 7.5. Biofuel production from microalgae.

Source: Image by Wikimedia Commons

Under nitrogen deficiency, certain microalgae can develop storage lipids (triglycerides) at up to 70% percent dry weight. *Chlorella* looks to be a viable feedstock for biodiesel synthesis among microalgae. When cultivated under heterotrophic conditions on glucose, *Chlorella protothecoides*, for example, yields a crude lipid content of 55.2 percent dry weight.

This alga's biodiesel has been demonstrated to be of a good grade, having a high viscosity and thermal efficiency.

Mass cultivation of microalgae not just to produces biomass for biodiesel manufacturing but also helps to mitigate the problem of global warming caused by carbon dioxide (CO2). Microalgae have substantially better photosynthetic efficiency than terrestrial plants and can flourish in high CO2 and temperature environments.

In principle, microalgae can utilize up to 9% of incoming solar energy to create 280 tonnes of dry biomass per ha^{-1} yr^{-1} while sequestering around 513 tonnes of CO2. *Spirulina* sp., *Scenedesmus obliquus*, and *Chlorella vulgaris* have been proven to grow at high CO2 levels and can grow up to 18 percent CO2. These microalgae are suitable for bioremediation of flue gas with 10–13% CO2.

Chlorella vulgaris cultivated in flue gas may fixate up to 4.4 g CO2 L^{-1} day^{-1}. When cultivated in a photobioreactor, *Chlorella vulgaris* can remove up to 74% of CO2 with a gas residence time of 2 seconds. As a result, an integrated system that uses microalgae for CO2 fixation and biodiesel production will be very much appealing.

7.8. CONCLUSION

Although macroalgae have received much attention as potential natural antioxidants, the value of microalgae as a source of natural antioxidants is further enhanced by the relative ease of purification of target compounds.

Some algae are considered rich sources of natural antioxidants. Algal biomass and algae-derived compounds have a very wide range of potential applications, from animal feed and aquaculture to human nutrition and health products. Highlight the role of dietary antioxidants and their potential benefits in health and disease directly or indirectly through plant nutrition and animal feed to produce healthy organic food.

Microalgae are photosynthetic microorganisms that are able to rapidly generate biomass from solar energy, CO2, and nutrients in bodies of water. Desirable chemicals are the products of secondary metabolism triggered under conditions not conducive to fast growth.

There is a current worldwide interest in finding new and safe antioxidants from natural sources such as plant material to prevent oxidative deterioration of food and to minimize oxidative damage to living cells.

This biomass consists of important primary metabolites such as sugars, oils, and lipids, for which process pathways exist for the production of high-value products, including human and animal feed supplements, transport fuels, industrial chemicals, and pharmaceuticals.

For those chemicals to be produced by microalgae, one needs to develop new strains (faster growth, higher substrate tolerance, etc.) by classical selection or genetic manipulation so microalgal biomass can be produced consistently. Reports on the antioxidant activity of microalgae are limited.

REFERENCES

1. Khan, A., 2018. Phytochemicals and their role in curing fatal diseases: A Review. Pure and Applied Biology, [online] 7(4). Available at: <http://www.thepab.org/files/2018/December-2018/PAB-MS-180100303.pdf> [Accessed 4 June 2022].
2. Kokou, F., Makridis, P., Kentouri, M. and Divanach, P., 2011. Antibacterial activity in microalgae cultures. Aquaculture Research, [online] 43(10), pp.1520-1527. Available at: <https://onlinelibrary.wiley.com/doi/10.1111/j.1365-2109.2011.02955.x> [Accessed 4 June 2022].
3. M. Amaro, H., Guedes, A. and Malcata, F., 2011. Antimicrobial activities of microalgae: An invited review. [online] Available at: <https://www.researchgate.net/publication/233820077_Antimicrobial_activities_of_microalgae_An_invited_review> [Accessed 4 June 2022].
4. S.M., S., 2012. Microalgal Biotechnology: Prospects and Applications. Plant Science, [online] Available at: <https://www.intechopen.com/chapters/41642> [Accessed 4 June 2022].
5. Sansone, C. and Brunet, C., 2019. Promises and Challenges of Microalgal Antioxidant Production. Antioxidants, [online] 8(7), p.199. Available at: <http://www.mdpi.com/journal/antioxidants> [Accessed 4 June 2022].

CHAPTER 8

NEGLECTED AND UNDERUTILIZED LEGUME CROPS: IMPROVEMENT AND FUTURE PROSPECTS

CONTENTS

There are numerous challenges that are impacting the world today, such as climate change, terrorism, and impecuniosity, which impede food production and availability, as well as ailment and nutritional security.

Ecumenically, food insecurity and low supply in many areas are threatening the human population and survival in the areas where terrorism and transborder and internal displacement of persons are entrenched in many components of the world.

8.1. INTRODUCTION

Food as a consequential commodity for survival is under threat, and if survival strategies are not devised, the catastrophe will be overbearing. A number of crop species are becoming extinct from our agricultural and forest fields, while some others are declining both in cultivation and utilization.

Sustainable food security denotes re-strategising crop genetic amelioration and engendering agronomy toward grain legumes to identify climate-resilient species and varieties with enhanced grain qualities.

This is highly desirable considering the paramount roles that grain legumes play in the food cultures around the world as veritable sources of quality protein, natural medicine, animal fodder, natural fertilizers, and environmental recuperation products, alongside the well-established soil enrichment property of symbiosis with nitrogen-fine-tuning bacteria.

It is crucial that alternative crops are cultivated and utilized to meet the ecumenical food requirements since they have been neglected and unexploited but have the potential to help increase food and dietary security in developing countries, particularly in sub-Saharan Africa.

With the recent negative impact of climate change being experienced worldwide, Africa is the most affected as a region that depends on rain-dependent agriculture. The effect of increased drought on crops has led to yield reductions at harvest, death of livestock, and loss of income and jobs in some components of Africa, especially in Somalia, Kenya, and Ethiopia.

This has drastically increased the effect of malnutrition and food insecurity. Most of the staple crops are unable to withstand the astringent environmental changes currently taking place. Nevertheless, Miscellaneous neglected and underutilized grain legumes (MNUGLs) are more resilient than the conventional staple crops.

Figure 8.1. Some Neglected Legume Crops.

Source: Image by PxHere

These MNUGLs are often linked to the cultural heritage of their places of inception, well acclimated to precise agroecological areas, adverse environments, and marginal lands. They withal perform well in traditional engendered systems with little or no external inputs.

Vegetable legumes have unique organoleptic qualities and are customarily regarded as consequential sources of carbohydrates, minerals, vitamins, and health-promoting bioactive compounds. In contrast, legumes are additionally known to carry some antinutritional factors like lectins, phytic acid, saponins, and vicine.

Moreover, legumes are characterized by their facility to develop in a symbiotic relationship with nitrogen-fine-tuning bacteria and ergo are withal utilized as soil-enriching green manure.

Some culled minor grain legumes with astronomically immense potential to boost protein security in periods of hunger and malnutrition in sub-Saharan Africa and elsewhere. The species are denoted by many names such as neglected, underutilized, underexploited, and under-researched, among others.

By incorporating recent advances in genomic technologies, these MNUGLs could be further enhanced to facilitate better acceptance and use, as well as to improve pabulum security.

8.1.1. Grain Legumes

The Fabaceae family of Angiosperms includes grain legumes, which are believed to be high in high-quality protein and are a significant influence on the nutrition, diet, and health of many people around the world. The family *Fabaceae* is divided into grain legumes and pasture/forage legumes.

The grain legumes are grown mainly as pulses providing food for humans, while pasture legumes are cultivated to feed domestic animals. Based on plant utility and economy, legumes are categorized into major and minor species. Major legumes are popular and common with well-established domestication and cultivation, agronomic practices, utilization, and conservation.

Grain legumes mainly consisting of common bean, pea, chickpea, faba bean, cowpea, lentil, pigeon pea, peanut, Asian Vigna species, grass pea, and horse gram are under cultivation in a considerable area worldwide.

With their higher protein content and symbiotic nitrogen-fixing bacteria in root nodules enabling them to fix their nitrogen, reducing fertilizer use in agriculture has become very important for the production systems. For most of these important grain legumes, a large number of germplasm accessions were characterized and evaluated for various agro-morphological traits, including biotic, abiotic, and quality parameters.

Core and mini-core collections have also been developed for the majority of grain legumes; they were further evaluated for different parameters. From these genetic resources, potential donors of desirable traits have been selected after evaluation and characterization and have been utilized in the genetic improvement of cultivars.

Current available genomic resources and technologies can facilitate allele mining for novel traits of interest and incorporation from wild relatives into elite domestic genetic backgrounds.

Examples include soybean (*Glycine max L.*), cowpea (*Vigna unguiculata L.*), groundnut (*Arachis hypogaea L.*), and common beans (*Phaseolus vulgaris L.*), pea (*Pisum sativum L.*), and chicken pea (*Cicer arietinum L.*), among others.

Minor legumes are less known, less exploited, neglected, and considered underutilized. Several species in this category include winged bean (*Psophocarpus tetragonolobus L.*), pigeon pea (*Cajanus cajan L.*), lablab (*Lablab purpureus L.*), lima bean (*Phaseolus lunatus L.*), jack and sword bean (*Canavalia sp.*), mung bean (*Vigna mungo L.*), Bambara groundnut (*Vigna subterranea L.*), Marama bean (*Tylosema esculentum L.*), Kersting's groundnut (*Kerstingiellageocarpa Harms*), African yam bean (AYB) (*Sphenostylisstenocarpa Harms*), and rice bean (*Vigna angularies L.*).

The wild species of the minor grain legumes include Kersting's groundnut (Kerstingiellageocarpa Harms), Marama bean (*Tylosema esculentum*), and the wild Vigna species such as *V. ambacensis, V. vexillata, V. luteola, V. oblongifolia,* and *V. racemosa*, among others.

These species are found in many African countries and could be exploited for food, medicine, agriculture (as to cover crops and fodder), and, more importantly, for genetic improvement of cowpea and related species.

Most of the MNUGLs have been relegated to unimportant underutilized crops grown by the older generation of farmers. Thus, sizable and valuable genetic resources housed within MNUGLs would have been lost due to neglect and lack of concerted, focused research.

Several authors have highlighted the usefulness of MNUGLs as food security in lean times as farmers in rural areas make a living on the species. Presently, there is no available genome sequence of MNUGLs species which could be utilized for successful breeding and specific purposes.

Some legumes are cultivated for their flowers, such as lupins, which are planted in gardens and are farmed commercially for their blooms.

Industrially farmed legumes include *Indigo fera* and *Acacia* species, which are cultivated for dye and natural gum production, respectively. Fallow or green manure legume species are cultivated to be tilled back into the soil to exploit the high levels of captured atmospheric nitrogen found in the roots of most legumes. Numerous legumes farmed for this purpose include *Leucaena, Cyamopsis*, and *Sesbania* species. Various legume species are farmed for timber production worldwide, including numerous *Acacia* species and *Castanospermum australe*.

8.2. SOME SELECTED MNUGLS

8.2.1. African yam bean (*Sphenostylisstenocarpa Ex. A. Rich Harms*)

The African yam bean (*Sphenostylisstenocarpa*) with somatic chromosome number 2n = 22 is a dicotyledonous species. AYB is an essential food crop in tropical Africa with great medicinal values and pesticide potential.

8.2.3. Winged Bean (*Psophocarpus tetragonolobus*)

The old legume *Psophocarpus tetragonolobus*, which has a 2n = 18 chromosome number, is commonly called 'winged bean.' The Winged bean is a multipurpose legume plant with all parts being edible and useful as medicine in the tropics of Asia, Africa, and Latin America.

Figure 8.4. A picture of the winged bean (*Psophocarpus tetragonolobus*).

Source: Image by Flickr

It is grown in many parts of the humid tropics, including Central and South America, the Caribbean, Africa, Oceania, and Asia. All parts of the plant are considered rich in vitamins, minerals, protein, and secondary metabolites such as phenolic and flavonoids. Leaves are usually eaten like spinach, and flowers are used in salads, tubers are eaten raw or cooked, while seeds are consumed when cooked.

In addition, winged beans are highly resistant to biotic and abiotic stresses and thus capable of growing under varying environmental conditions. It is now the favorite specimen of many scientists trying to explore its rich potential.

The seeds of winged beans also exhibit tolerance to storage pests. Apart from its seeds, the tubers/roots are also nutritious and rich in protein, about 20%, while the leaves and flowers are also high in protein (10–15%).

8.2.4. Lima Bean (*Phaseolus lunatus L.*)

Throughout the Caribbean, Peru, Mexico, and Asia, the lima bean is grown for its edible seeds and as a leafy vegetable. Rich in protein, lima beans are resistant to viral and rust diseases and withstand insect pests, drought, and abiotic stress. The species also tolerate different levels of aluminum and manganese toxicity which can be exploited to advance the sustainable utilization of other legumes.

Figure 8.5. A picture of Lima beans (*Phaseolus lunatus L.*).

Source: Image by PixaHive

8.2.5. Hyacinth Bean (*Lablab purpureus L.*)

Hyacinth bean (*Lablab purpureus L.*) is cultivated for its edible seeds and pods. It is mainly grown in Africa and Asia as a source of food in the form of vegetables, green pods, and seeds. Several field trials suggested that the species is drought tolerant and water-efficient, and produces a high yield.

The protein content is comparable to that of soybean. Reports also indicate that *L. purpureus* has the potential to be a source of pharmaceuticals and nutraceuticals as medicine and traditional medicine in Asia and Africa.

Figure 8.6. A picture of the Hyacinth bean (*Lablab purpureus L.*).

Source: Image by Flickr

8.2.6. Jack Bean (*Canavalia ensiformis L.*)

As the most economically important species of the genus *Canavalia, Canavalia ensiformis* is also known as jack bean. Jack bean has enormous potential as a food source for humans and livestock alike. It is widely distributed in Africa, Asia, and America, with large-scale cultivation reported in Congo and Angola.

It is rich in protein and thrives well in poor and acidic soils. Jack bean is mainly grown for its nutritious pods, seeds, and fodder. It is a forage crop with a high green manure capacity to enrich the soils and also to control soil erosion.

Figure 8.7. A picture of Jack beans (*Canavalia ensiformis L.*).

Source: Image by Wikimedia Commons

The crop tolerates adverse environments, drought, heat, and leached soils; also, it resists pest attacks. The leaf of the jack bean contains crude proteins and fiber compared to other legumes. Jack bean possesses a deep root system that enables the plant to penetrate deeply into the soil, which enables it to withstand very dry conditions.

Raw jack beans contain toxic compounds such as tannin, phytate, saponins, canavanine, concanavalin A (hemagglutinin), and trypsin inhibitors.

Sword bean (*Canavalia gladiata L.*) is another species in the *Canavalia* genus of the *Fabaceae* family that has many uses, including as a food, leafy vegetable, medicine, forage, and a cover crop.

It is a vigorous perennial climber plant usually cultivated as an annual. Reports indicated that the sword bean originated from the Asian continent and is now known in the tropics as an introduced species. The red sword bean is one of the edible beans of China reportedly rich in antioxidant polyphenols with great medicinal uses.

Furthermore, the seed coat of the bean is rich in gallic acid and its derivatives, mainly gallotannins, a common trait found in legume polyphenols. The chemical composition of seeds of sword bean has been reported and compares quite well with soybean.

The average yield ranges from 720 to 1500 kg/ha, which can be compared with a soybean yield of 600–1000 kg/ha. The fruits mature in 6–10 months after planting. The sword bean is relatively resistant to attack from pests and diseases.

8.2.7. Pigeon Pea (*Cajanus cajan L. Millsp*)

Cajanus cajan, also known as pigeon pea, is an erect perennial shrub, or woody plant, that is widely cultivated in tropical regions. Pigeon pea is mainly cultivated for its edible seed grains as well as feed, forage, and fuel. It has a diploid genome with a somatic chromosome number of $2n = 22$.

Most farmers depend on *C. cajan* as an alternative source of protein to support workers and families during lean times. The plant grows well in areas with low rainfall and varying climatic conditions. It is a drought-tolerant crop capable of withstanding poor soil and abiotic stress. Diversity exists in seed coat color, size, texture, and taste.

Moreover, the grain protein content of cereal crops following a legume crop is increased, and the soil structure and health are improved after growing legumes. Accordingly, nitrogen-fixing legumes provide unparalleled sustainable opportunities for minimizing future nitrogen fertilizer usage.

The inclusion of grain legumes in cropping systems can increase cropping intensity, enhancing annual productivity as well as increasing diversity and reducing overall risk because the reliance on one or more crops is decreased.

While legume crops can favor soil acidification due to an imbalance in cation over anion uptake, this can be managed by varying the legume species used in the crop rotations and by the application of lime in the soil. However, the sustainable development of intensive agriculture can be affected when grain legumes are intensively grown on the same farmland year after year, for instance, because auto toxins build up and affect soil microbe communities.

8.4. MISCELLANEOUS, NEGLECTED AND UNDERUTILIZED GRAIN LEGUMES (MNUGLS)

There are many different ways of identifying neglected and underutilized species (NUS), including their usefulness in the local context, localized domestication, adaptation coupled with a general abandonment by mainstream agricultural researchers, extension services, plant breeders, donors, technology providers, policymakers, among others. NUS are classically identified based on certain criteria, which include the following:

- Local importance in consumption and production systems
- Adaptation to agroecological niches/marginal areas
- Representation by ecotypes/landraces
- Cultivation and utilization based only on indigenous knowledge
- Rare representation in ex-situ collections
- Uncoordinated attention from national agricultural and biodiversity conservation policies, research, and development
- Neglect by the mainstream market system

In reality and broad consideration, a large percentage of such NUS are underutilized legume crops. The species of these legumes are classified as minor grain legumes even though they are widely cultivated as food and forage crops. Thus, we can refer to this category of legume crops as MNUGLs.

On global distribution, these species are endemic to the tropical regions of the world. Several reports and findings have established that MNUGLs are drought-tolerant, endure and thrive under harsh environments, are highly adaptable to varying ecogeographical settings, and withstand or mitigate conditions such as heat, drought, diseases, frost, cold, and insect pest attack.

These qualities could be scientifically explored for crop improvement and sustainable utilization. Aside from these good qualities, MNUGLs are also considered high-quality legumes because they contain high levels of protein and micronutrients comparable to those found in conventional legumes. They are also indispensable in crop rotation strategies to fertilize agricultural soils.

8.5. YIELD STABILITY IN GRAIN LEGUMES

Grain legume yields are considered to be more variable than those of cereals, and yield fluctuations are one of the main reasons farmers give for not growing these crops. However, little and contradictory quantitative evidence is available on yield stability in grain legumes.

Some statistical analyses suggest that the yield stability of grain legumes is higher than that of other crops, whereas others indicate the lowest yield stability for grain legumes across Europe and the United States. Pea was the most variable in four, lupin in four, rapeseed in five, and small-grain cereals in one, indicating that broad-leaved crops indeed have more unstable yields than cereals.

However, statistical analyses of yield stability using the coefficient of variation have several weaknesses. Few of the available datasets are spatially explicit, and grain legumes that are often grown on less productive soils were compared with crops grown on all types of soils, which would negatively influence yield variation statistics. Hence, other stability parameters are needed.

8.6. MITIGATING THE IMPACT OF CLIMATE CHANGE

Future legume production will be influenced by climate change factors such as:

1) Increased atmospheric CO2 levels favoring carbon gain because most legumes use C3 photosynthesis;

2) Faster plant developmental rates due to the predicted higher temperatures, which would allow a shorter growing season and reduce exposure to the drought that is often experienced at the end of the cropping season
3) Accelerated canopy decay due to extreme temperatures;
4) Reduced photosynthetic efficiency, increased pod and flower abortion, and reduced production of reproductive structures due to more frequent droughts;
5) Defective pollination due to high temperature-induced pollen sterility; and
6) Reduced seed quality. Regional yield changes will depend on the local manifestation of climate change (the interaction between elevated CO2, high temperature, and drought) as well as other factors.

Global soybean production already appears to have been influenced by the negative impacts of climate change. Overall, soybean yields declined by 2–4% for every degree rise in temperature during the growing season in the USA between 1994 and 2013, resulting in losses of US$11 billion.

Rising global temperatures could exacerbate heat stress and reduce the areas suitable for bean production. Future climate conditions are predicted to be more favorable to common bean cultivation in the Northern Hemisphere than in the Southern Hemisphere; new grain legume growing areas should open up in Canada, northern Europe, and Russia.

For example, pea and faba beans are becoming increasingly important crops in Finland. Adaptations to climate change in developed countries, such as Canada and France, will increase yields of dry peas, chickpeas, broad beans, lentils, lupins, and grass peas. However, yields are expected to decrease in developing countries in the tropics and sub-tropics, such as India, Pakistan, and Ethiopia.

There is a yield gap of more than 3000% among legume crops in Africa; cowpea yields are only 10–20% of their genetic potential. Moreover, while legume cultivation is an integral part of Indian agriculture, legume crop production has remained low and unstable, with a yield gap for soybean ranging from 850 to 1320 kg/ha, for groundnut 1180 to 2010 kg/ha, for pigeon pea 550 to 770 kg/ha and 610 to 1150 kg/ha for chickpea.

Through improved crop management, it will be possible to improve productivity by 800 kg/ha on average for major grain legumes.

8.7. RESEARCH EFFORTS AND CONSTRAINTS TO THE GLOBAL CULTIVATION AND ADOPTION OF MNUGLS

MNUGLs are increasingly becoming rare across the world, with their associated valuable genetic resources disappearing rapidly in all their natural ranges. Therefore, there is the need for a paradigm shift from the present scenario of neglect to sustainable cultivation, exploitation, and utilization of the species.

In recent years, grain legume stakeholders have advocated for increased global cultivation and production of MNUGLs toward a sustainable solution to food and protein security, plus agricultural and environmental restoration.

As well, some efforts are being undertaken to improve legume crops' nutritional quality and yield, as well as enhance their resilience to climate change. This is done through enhanced genomic resources and breeding techniques.

Across a wide range of food systems and agroecological niches, MNUGLs are highly adaptable to agroecological niches/peripheral regions. This contributes substantially to global protein security, beneficial agricultural practices, and alleviates rural poverty, among others.

To advance food and nutritional security in the future, as well as maintain biodiversity and mitigate the dangerous effects of climate change and abiotic stress, the potential genetic resources at and within the MNUGLs have not been explored.

In addition, the capacity of species to withstand abiotic stress even in the face of biologically limiting factors is also important for continuing exploitation and survival. For MNUGLs to be exploited for present and future food, nutrition, and protein security, a holistic approach is necessary to prevent total genetic erosion.

Growing and using MNUGLs sustainably is hindered by the long cooking time of seeds, the growth habit that requires staking, the large labor requirements, and the lack of staking materials. In addition, poor seed quality, high labor costs, postharvest diseases, and antinutritional factors are all factors that contribute to poor crop performance.

The development of breeding intervention approaches, constraints, and possible solutions to the cultivation and utilization of MNUGLs.

8.8. GENETIC POTENTIALS OF MNUGLS

In recent times, the Food and Agriculture Organization (FAO) of the United Nations estimated that around 800 million people, particularly in developing countries, suffer from food and nutrition insecurity.

The Sustainable Development Goals (SDGs) of the 2030 Agenda adopted by the United Nations in September 2015 also aimed at having zero hunger as one of its SDGs by abolishing hunger and malnutrition, especially in the less developed world.

To deal with this challenge, a tactical approach would be to promote biodiversity and use of neglected and underutilized crop species in people's diets and food patterns. The MNUGLs would increase food production levels, diversify human diets, and enhance the sustainable utilization of a variety of climate-smart crops.

In addition, MNUGLs contain nutrients and health-enhancing compounds capable of preventing malnutrition and some chronic diseases. Educating local communities about the many health and nutritional benefits of MNUGLs could also prove to be an effective tool in improving human nutrition. To meet human nutritional needs in developing countries, MNUGLs could be exploited to their full potential.

Even though MNUGLs possess enormous genetic potential, genetic erosion or loss of genetic resources is alarming. Researchers should focus on identifying and utilizing genetic resources within the minor crop gene pool. To prevent the continuous loss of genetic resources among the MNUGLs, sustained research efforts are needed.

MNUGLs face a challenge in developing effective phenotyping and breeding methods, according to recent reports. A lack of genetic diversity limits efforts to improve disease resistance, quality, and yield in modern breeding programs. Gene banks contain several varieties of grain legume seeds, but these varieties have not fully been incorporated into current breeding programs.

A large quantity of these minor grain legumes is available in Africa, which can mitigate the effects of malnutrition and poverty in sub-Saharan Africa if properly harnessed. The diversity existing among the seeds of the species is worthy of research attention for food, agriculture, and medicine.

8.8.1. Genetic Improvement and Prospects of MNUGL

Genetic and breeding efforts to improve the underutilized and neglected legume crops in architecture, period of maturation, yield, and nutritional contents have not recorded commiserate level of success as expected. Traditional hybridization and other breeding techniques, though they have been used for some desired intentions, are yet to translate to desired results.

Few successful crosses have been recorded so far on MNUGLs. Few successes have been reported on *Cajanus cajan* and some species. Reproductive barriers such as embryo abortion have been reported in many MNUGLs as limiting factors to genetic enhancement/improvement. However, tissue culture and micropropagation could be further employed to overcome such barriers with the capacity to generate fertile haploid plants.

Advancement in DNA technology has enhanced our understanding of the huge pote available in the genome of many plant species and particularly the underutilized legumes. Several genomic breakthroughs involving the genetic engineering of cereal crops have been reported.

DNA-based methods are reliable and have been employed to identify, trace, and certify plant genealogies, origins, and phylogenetic relationships. DNA barcoding has been applied to identify and characterize some underutilized and neglected legumes such as *Lablab purpureus, Tylosema esculentum, Vigna subterranean, V. vexillata,* and *Vigna unguiculata.*

The use of molecular markers for marker-assisted selection (MAS) or breeding programs has played a significant role in the assessment of the level of genetic diversity or relatedness among various species of underutilized legumes.

Globally, different DNA techniques such as random amplified polymorphic DNA (RAPD), amplified fragment length polymorphisms (AFLPs), inter simple sequence repeats (ISSR), single feature polymorphisms (SFP), single-nucleotide polymorphisms (SNP), and chloroplast gene RBCL, among others, have been employed to evaluate the genetic relationships and diversities among neglected underutilized species.

RAPD and ISSR markers were used for genetic diversity studies in winged beans by Mohanty et al.

The physiological and phytochemical parameters of the genotypes were investigated. Distinct winged bean novel lines were identified, and the information from the analysis of photosynthesis rate, photosynthetic yield, and stomatal conductance data revealed two clusters in correspondence with the phytochemical affinities of the genotypes.

The use of SSR, SNP, and genotype by sequencing (GBS) for the studies of phylogenetic relationships and genetic diversities among the MNUGLs is rare due to a lack of sequence information; hence RAPD and AFLP were used for such species, and attempts were made to transferability of specific SSR markers in cowpea for genetic diversity studies in underutilized legumes by other authors, as in the case of African yam bean.

Globally, the complexity of plant genomes has led to advancement in genome sequencing, determination of polyploidy, genome size, repetitive DNA sequences, and transposable elements toward genetic engineering of plants to generate useful products apart from innate uses. Transposable elements (TEs) are ubiquitous in flowering plant genomes, of which a higher percentage of such genomes are occupied by us.

TEs, via their amplification, methylation, and recombination, contribute to the restructuring of plant genomes, epigenomes, centromeric regions, and the evolution of new genes for novel genetic functions.

Identification of TEs in a species is critically significant to the understanding of their functional roles. Therefore, a detailed description of this is a major procedure to precisely identify specific genes and evaluate the association between genes and TEs in a complex sequenced genome.

Such studies have not been applied to MNUGLs to understand the role of transposons in the long duration of seed cooking of most of the MNUGLs and the expressivity of secondary nutritional metabolites.

We believe that such studies will have an overall influence on the genetic manipulation of the MNUGLs, understanding of potential gene-TE interaction, identification of active TEs for functional genomics, and development of TE-based molecular markers for genotyping studies.

Currently, sophisticated sequencing genomic approaches such as de novo transcriptome sequencing are being utilized to identify and describe key genes responsible for varied economic, nutritional, physiological, and pharmaceutical uses of plant species. Transcriptome sequence analysis is one of the molecular tools that can also be applied to MNUGLs for improvement purposes.

It is hoped that some of these tools will be employed in due course, not only to analyze genetic diversity among the MNUGLs but also to identify key genes that will be potentially useful for breeding and utilization purposes.

Genes that are useful for varied needs could be identified, described, and extracted from the MNUGLs, thus ensuring the sustainable utilization of the species. Proteomics and genomics are increasingly being applied to unravel several genetic constraints and proffer robust solutions toward their sustainable production and utilization.

These areas, combined with metabolomics, offer great possibilities in the quest for the improvement of MNUGLs. Consequently, the application of molecular breeding tools such as marker-assisted selection, genomic selection (GS), and genome-wide association (GWAS) has been appraised to influence scientific efforts for improving the grain yield of orphan crops in developing countries.

These, of course, including next-generation sequencing (NGS), have greatly enhanced the improvement of many commercial crops, which the MNUGLs can also benefit from.

Recent advancement in omics technologies such as genomics, proteomics, transcriptomics, and metabolomics has equally enhanced our understanding of the genetic structure of plant species, as well as the expression of genes through transcriptomic/proteomic profiling and their role in the overall metabolism of plants.

Vatanparast etal. used transcriptome sequencing to develop SSR and SNP markers for winged bean (*Psophocarpus tetragonolobus*) and also gave insights into the divergence of the Kunitz-type trypsin inhibitors, which are essential antinutritional factors in winged bean and other legumes.

Transcriptome sequencing is an inexpensive and reliable method for efficient and rapid identification of molecular markers in underutilized plant species.

Prospects also lie in the adoption of high-throughput tools, including gene editing, GWAS, and clustered regularly interspaced short palindromic repeat (CRISPR), among others, toward the genetic improvement of the species for sustainable cultivation, production, and utilization.

8.9. NEW TECHNOLOGIES FOR LEGUME IMPROVEMENT

Grain legumes have underpinned the development of genetics. The common garden pea was used by Gregor Mendel to demonstrate the 'particulate nature of inheritance' in 1865. By and by, many grain vegetable reproducing programs experience the ill effects of low hereditary variety and low paces of hereditary advancement.

Chickpea encountered a few bottlenecks during and after taming and thusly has restricted hereditary variety in the yield genetic supply and on account of soybean, five presentations represented 55% of the family in open soybean cultivars in the USA during the 1990s.

Hereditary improvement of peas was sped up by early age choice in light of the creature model. Such strategies could add to the practical genetic improvement of the world's grain vegetable harvests and help to overcome any barrier between grain vegetables and their wild and landrace family members.

When joined with genomics-assisted breeding, it will be feasible to transfer important genes, for example, resilience to drought stress in wild vegetables, and transfer them productively into developed varieties.

Fast introgression of these significant gene is a significant test confronting grain vegetable reproducers, who need to speed up grain vegetable hereditary improvement to measure up to the assumptions of the 2009 Declaration of the World Summit on Food Security, which expressed that a 70% increase in farming produce was expected by 2050 to match with populace increase while adjusting to environmental change through maintainable utilization of hereditary assets for food and farming.

The significance of biodiversity in seed banks has been perceived, and much consideration has been paid to rice and other oat crops. To advance and upgrade the development of vegetables, germplasm varieties are significant genetic assets, and enormous genetic and phenotypic varieties exist on the planet.

It is consequently vital to have an efficient stock of landraces and varieties. The vast majority of the freely accessible data can be found in GENESYS (Global Portal on Plant Genetic Resources. Notwithstanding, major CGIAR Institutes recorded in critical genes of grain vegetable germplasm are rationed in different public genetic databases sites.

Germplasm from China can be gotten to using the Chinese Crop Germplasm Resources Information System and the Crop Germplasm Resources Platform under the Ministry of Science and Technology, P.R. China.

The National Institute of Agrobiologically Sciences (NIAS) GenBank holds the biggest germplasm database in Japan. The germplasm from India can be obtained through the National Bureau of Plant Genetic Resources (NBPGR) database.

This germplasm list isn't comprehensive since data is frequently difficult to recover, and a few increases are copied across hereditary asset habitats. Additionally, the arrangement of the information could be normalized to work with the clients.

Entire genome sequencing has turned into a reasonable and useful asset to depict genomic data of germplasm. Genomic data can be utilized to produce high-goal hereditary information for significant agronomic attributes, foster markers for reproducing, and distinguish significant genes for crop improvement.

High-goal hereditary information are accessible for 10 vegetables with new grouping data, and low-goal maps are accessible for everything except Bambara bean, tepary bean, and lima bean. The foundation of hereditary assets and grain vegetable genome sequencing together sped up the advancement of genomics-assisted breeding systems toward crop improvement.

Worldwide crop iprovement has nearly significantly increased throughout recent years; however, grain vegetable improvement has just expanded by around 60%. The somewhat low pace of yield improvement in grain vegetables versus cereals is part of the way made sense of by low hereditary variety in grain vegetable breeding projects.

It is critical to increment hereditary variety in first-class rearing projects, assuming we are to benefit from biotechnology for vegetable yield improvement. The genomic choice depends on allelic variety in the reproducing populace for complex genes and may work on long haul hereditary advancement assuming joined by large successful populace size with negligible inbreeding.

Determination for complex characteristics was demonstrated to be more proficient when given genomic relationship data in plants. For grain vegetables, a considerable lot of which are self-pollinating crops, genomic

genome. Utilizing a mix of approaches, more than 28 K non-repetitive gene models were anticipated across the chickpea genome, of which around 25 K (89.73%) could be practically commented on.

Alongside the chickpea genome, sequence information was created for 90 chickpea promoters utilizing Whole-genome re-sequencing (WGRS) and RAD-sequencing approaches. In one more work to group the chickpea genome, the ICC 4958 chickpea genotype was focused on creating a draft genome gathering utilizing NGS stages alongside BAC-end sequences and a hereditary guide.

A sum of 13.35 Gb great sequencing information from 454/Roche GS FLX Titanium stage and 43.7 Gb of value separated PE Illumina sequence information on ICC 4958 alongside BAC-end sequences were utilized to collect ~520 Mb of chickpea genome.

Recently, an improved version of *desi* chickpea cultivar ICC 4958 with 2.7-fold increase in the length of pseudomolecules was reported. This improved assembly could reduce the gaps in the current genome assembly and also, predicted the presence of more than 30 K protein-coding genes.

In a similar effort, used method of chromosomal sequencing to validate the available *desi* and kabuli chickpea genome assemblies. Isolation and NGS based sequencing of individual chromosomes aided in validating the genome assemblies at chromosome level, which identified small misassembled regions in kabuli and large misassembled region in desi draft genomes.

As of late, the International Peanut Genome Initiative (IPGI) effectively sequenced the genomes of diploid groundnut begetters. The forebears addressing A-genome (*Arachis duranensis*, promotion V14167) and B-genome (*Arachis parenesis*, increase K30076) together address the tetraploid genome of developed groundnut (*Arachis hypogaea*).

So, in a similar context, a total of 216 Gb WGS data at 154.4X coverage for accession V14167 whereas 168.8 Gb data at 120.6X coverage for accession K30076 were generated. In addition to that, 155.5 Mb transcriptomic sequence data for accession V14167 and 175.6 Mb data for accession K30076 were also generated from different plant tissues and growth stages.

The sequences mentioned above were assembled into 10 pseudomolecules for each genome. The B-genome (1.37 Gb) had larger genome size than A-genome (1.2 Gb). The availability of the genome sequence will provide access to 97% of groundnut genes in their genomic context to the global groundnut research community which leads to development of better understanding regarding the complex groundnut genome and accelerated development of more productive, climate, and stress-resilient groundnut varieties.

Legumes have been included in cropping systems for hundreds of years, especially in rotation with other crop species. A boost in grain legume production is urgently needed to oppose the static or declining production trends especially in developing countries, despite increasing global demand. The worldwide year of pulses (grain legumes) in 2016 gives an amazing open the door to ponder the situation with worldwide grain vegetable creation, utilization and possibly open doors for future development. Our ongoing over-reliance on a small group of significant staple crops has inborn agronomic, natural, healthful, and financial dangers and cheapens the commitments made by under-used vegetables.

Admission of a different group of vegetables is significant in the human eating regimen. Besides, numerous underutilized vegetable crops are now a fundamental source of nutrients, micronutrients, and protein for production and, consequently, an important part of food security.

Habitually, underutilized vegetable yields are not generally developed as a product but rather gathered from unchanged landraces or wild populaces. The innate versatility and stress resistances of these nutritious and energy-adequate species could assume a significant part in guaranteeing food security even with environmental change for populaces living in underestimated areas.

The exchange of innovations connected with grain vegetable creation to asset-poor country networks could be hurried with more participatory approaches.

The ongoing utilization of nitrogen composts in horticulture is ~110 million tons dad, with the greater part coordinated to oat creation. Just 30-half of applied nitrogen is utilized by the planned harvest, and the abundance of nitrogen preparation adversely affects environmental change and biodiversity.

The capacity of vegetables to fix air nitrogen through advantageous interaction with rhizobia has a tremendous generally undiscovered possibility for manageable horticulture, plant variety, and improvement of essential creation with diminished manure use, helps which may likewise stretch out to phosphorus-poor soils.

Late expanding nitrogen compost costs are focusing more consideration on further developing productivity in editing systems. Endeavors to make nitrogen-fixing cereals are in progress, yet grain vegetables right now get less innovative work consideration.

Vegetables are behind grains as far as regional development and efficiency gains, notwithstanding expanding worldwide interest. This slack might be expected to some extent to unsound grain vegetable costs because of high fluctuation in their yields and competition from high yielding oat crops.

What's more, government cost help systems exist, especially in non-industrial nations, as well as deficient contributions to reproducing and agronomic advancements expected to further develop yields.

8.10. CONCLUSION

Most produced vegetables utilized for food are devoured as grain seeds known as pulses. In any case, the development of some vegetable species is focused on their utilization .

According to a well-being viewpoint, vegetables are viewed as important sources of plant protein, carbs, essential minerals, nutrients, and phytochemicals. Significantly, vegetables contain low-fat levels with great levels of edible proteins.

Vegetables have a large variety and different stress resistance capacities; they could upgrade food and food security in low-pay areas of Africa. Investigating underexploited vegetables' healthy benefits can beat hunger, particularly in Sub-Saharan Africa, in non-industrial countries.

Essentially, supplement strength can be broken down and utilized in plans for the food-to-food approach. Wild family members of underexploited vegetables for subjective and quantitative attributes and their domestication will contribute fundamentally to food security.

Normal yields of vegetables have changed altogether in late many years. This accomplishment was expected to some extent, to the reproduction of better-performing crossovers created by joining CWR's ingrained plant genomes.

In addition, ongoing genomic advances and accessibility of vegetable genome information speeds up the variety development. Notwithstanding, to help the world's rising populace, supporting the pace of expansion in vegetable production will be fundamental.

This challenge will most likely be tended to through better cultivating, substantially more dependable seed supplies, in addition to additional steady business sectors also by the use of genomics. Most financially significant plant groups rely upon the cooperation of huge quantities of genes.

With genomics, breeders can describe the allelic attributes of their specific germplasm in the breeding system and hence generate valuable allelic blends.

A mix of genomics apparatuses with traditional techniques would help breeders in their endeavors to plan and appropriately choose the best blends of genes and alleles in the linked species to satisfy the necessities for crop creation advancements.

REFERENCES

1. Bhattacharya, A., 2019. Nitrogen-Use Efficiency Under Changing Climatic Conditions. *Changing Climate and Resource Use Efficiency in Plants*, [online] pp.181-240. Available at: <https://www.sciencedirect.com/science/article/pii/B9780128162095000040> [Accessed 4 June 2022].
2. Pandey, M., Roorkiwal, M., Singh, V., Ramalingam, A., Kudapa, H., Thudi, M., Chitikineni, A., Rathore, A. and Varshney, R., 2016. Emerging Genomic Tools for Legume Breeding: Current Status and Future Prospects. *Frontiers in Plant Science*, [online] 7. Available at: <https://www.frontiersin.org/articles/10.3389/fpls.2016.00455/full> [Accessed 4 June 2022].
3. Popoola, J., Ojuederie, O., Omonhinmin, C. and Adegbite, A., 2020. Neglected and Underutilized Legume Crops: Improvement and Future Prospects. *Recent Advances in Grain Crops Research*, [online] Available at: <https://www.intechopen.com/chapters/68218> [Accessed 4 June 2022].
4. Shafique, A., Rehman, S., Khan, A. and Kazi, A., 2014. Improvement of Legume Crop Production Under Environmental Stresses Through Biotechnological Intervention. *Emerging Technologies and Management of Crop Stress Tolerance*, [online] pp.1-22. Available at: <https://www.sciencedirect.com/science/article/pii/B9780128008751000016> [Accessed 4 June 2022].
5. Stagnari, F., Maggio, A., Galieni, A. and Pisante, M., 2017. Multiple benefits of legumes for agriculture sustainability: an overview. *Chemical and Biological Technologies in Agriculture*, [online] 4(1). Available at: <https://chembioagro.springeropen.com/articles/10.1186/s40538-016-0085-1> [Accessed 4 June 2022].
6. Wilson, R., 2019. Legume Crop Genomics. [online] Available at: <https://www.taylorfrancis.com/books/mono/10.1201/9780429103933/legume-crop-genomics-richard-wilson> [Accessed 4 June 2022].

INDEX